Microwave and cellular comm planning and design

For engineers and managers

Second edition

Abdiasis Abdallah Jama

©2021, 2020

ALL RIGHTS RESERVED. No part of this publication may be reproduced, stored in a retrieval system, or transmitted, in any form or by any means, electronic, mechanical, photocopying, recording, or otherwise, without the prior written permission of the author. Published in Somalia.

Dedication

To my mother Ruqia Hajji Warsame

Contents

1. Preface 6
2. Acronyms 7
3. Fundamentals of electromagnetic waves theory 9
   Digital communication link performance 23
4. Introduction to Microwave transmission 25
5. Project management 33
6. Route design 36
7. Link design and installation 47
Link configuration and installation 47
Link traffic configuration and protection 63
Networking and protocols 77
Satellite links 90
8. Frequency and capacity planning 104
9. Interference analysis 114
10. Telecom regulation and spectrum monitoring 119
11. Cellular network planning fundamentals 128
Introduction 128
GSM 136
UMTS 144
LTE 148
Interoperability 166
   Intra RAT cell reselection 167
   Inter RAT cell reselection 174
   Intra RAT handover 180
   Inter RAT handover 188
   Network optimization 196
Introduction to 5G NR R15 213
12. Appendix A: Cellular FDD RF channels 233
13. Appendix B: NR FDD FR1 bands (138 104 v15.4.0 table 5.2-1) 235
14. Appendix C: NR TDD FR1 bands (138 104 v15.4.0 table 5.2-2) 236
15. Appendix D: NR transmission bandwidth configuration (TS 138 104 V15.4.0) 237

16. Appendix E: EN-DC operating bands..238
17. Appendix F: Microwave channel arrangement..239

Preface

Mobile Communication industry is evolving and so is the required knowledge and experience of planners and managers. In mobile communication the access network is the part that is the most challenging to the operator in terms of significant investment it demands both in CAPEX and most importantly in OPEX. Therefore capacity building in this area is what makes operators competitive with increased customer reputation

The microwave transmission network is the backbone and backhaul part of the network. It needs deep understanding of electromagnetic propagation, fading, redundancy route and link design, frequency and interference management, payload migration from TDM to Ethernet to IP, and satellite backup link in order to achieve 99.999% performance objectives. The microwave industry is evolving to compete with fiber to meet with 4G and the newer 5G high capacity needs. The objective of this book is to guide engineers and managers to practical planning, designing, implementing, and maintaining microwave transmission networks that can meet requirements of 3G to 5G systems. Best installation practices that lead to high performance transmission link is also addressed

Cellular network has also evolved from 1G to 5G. In the past coverage and capacity were planned separately. With 3G and 4G coverage, capacity, and interference layers must be merged together to produce reliable results. This book will walk you through cellular system from 2G to 4G planning in coverage, capacity, cell RF parameters, interoperability, and KPI optimization from first principles

Past experience of the author has shown that quality of cellular system is dependent on quality of microwave backhaul links. Engineers and managers need to understand both microwave and cellular in order to optimize the overall wireless network and minimize OPEX

Having that said engineers and managers who read this book will be able to plan and design both microwave and cellular networks in high quality standard. It is therefore the ultimate manual that will help you be competitive in your technical work

Acronyms

EIRP	Effective isotropic radiated power
QAM	Quadrature amplitude modulation
RF	Radio frequency
ODU	Outdoor unit
DEM	Digital elevation map
GPS	Global positioning system
SRTM	Shuttle radio topography mission
LOS	Line of sight
dB	Decibel scale
VLAN	Virtual local area network
STP	Spanning tree protocol
MPLS	Multi-protocol label switching
OSPF	Open shortest path first
BTS	Base station sub-system
UE	User equipment
GSM	Global system for mobile communication
GPRS	General packet radio service
UMTS	Universal mobile terrestrial system
WCDMA	Wideband code division multiple access
LTE	Long term evolution
RNC	Radio network controller
BSC	Base station controller
MME	Mobility management entity
MSC	Mobile switching center
HSS	Home subscriber server
VLR	Visitor location register
SGW	Serving gateway
PGW	Packet gateway
PCI	Physical cell identifier
PSC	Primary scrambling code
BSIC	Base station identifier code
TAC	Tracking area code
LAC	Location area code
RAC	Routing area code
PRACH	Physical random access channel
BCCH	Broadcast control channel
CPICH	Common pilot control channel
RS	Reference signal
RSSI	Receive signal strength indicator
RSCP	Receive signal code power
RSRP	Receive signal reference power
RSRQ	Receive signal reference quality
SINR	Signal to interference noise ration
CSFB	Circuit switched fallback
NCC	Network color code

Acronyms

BCC	Base station color code
PSS	Primary synchronization signal
SSS	Secondary synchronization signal
ARFCN	Absolute radio frequency channel number
MCC	Mobile country code
MNC	Mobile network code
PLMN	Public land mobile network
CGI	Global cell identifier
MAIO	Mobile allocation index offset
HSN	Hopping sequence number
TCH	Traffic channel
SDCCH	Slow associated control channel
SIB	System information message
MIB	Master information block
HSPA	High speed packet access
PUSCH	Physical uplink shared channel
PUCCH	Physical uplink downlink channel
PDSCH	Physical downlink shared channel
RRC	Radio resource control
C-RNTI	Cell radio network temporary identifier
VSWR	Voltage wave standing ratio
ECFM	Ethernet connectivity fault management
ERPS	Ethernet ring protection switching
VPN	Virtual private network
GEO	Geo stationary orbit
MEO	Medium Earth orbit
UL	Uplink
DL	Downlink
LNB	Low noise block converter
BUC	Block up converter
LO	Local oscillator
ITU	International telecommunication union
ETSI	European telecommunication standards institute
ANSI	American national standards institute
CCDP	Co-channel dual polarized
XPD	Cross polarized discrimination ratio
XPIC	Cross polarization interference cancelling
RSL	Receive signal level
IEEE	Institute of electrical and electronic engineers
3GPP	Third generation partnership project
CFRA	Contention free random access
SSB	SS-Block

Chapter One

Fundamentals of electromagnetic waves theory

All wireless communication whether radio, cellular, microwave, satellite, and even wired communication such as optical carry the user information on electromagnetic wave. In the wireless case the antenna radiates the electromagnetic wave in either directional or Omni directional. Microwave and Satellite transmission antenna system radiate directional while cellular system use Omni directional antenna

So before we go into more practical transmission and cellular network planning and operators, let us take view minutes in physics classroom and explain intuitively the electromagnet wave from first principles of electric and magnetic fields. Telecom managers may feel bored and confused by reading this chapter as they want more practical rather than theoretical background. But they can skip to next chapter though the author recommends to read this first chapter. It has been simplified.

In electromagnetics we are dealing with electric and magnetic fields. We can study them in the following ways
- Static electric fields which does not vary with time. These fields are produced by point charges and charge distributions
- Static magnetic field which does not vary with time. These fields are produced by steady current.
- Time varying electromagnetic field that is produced by time varying currents. In this case time varying electric field produces magnetic field (Ampere's law) and time varying magnetic fields produce electric fields (Faradays law)

For static electric fields we have the following electrostatic rule

Gauss law states that:

Electric flux passing through a closed surface is equal to total charge enclosed in the surface.

$$\int D.ds = Q \qquad (1)$$

D is the electric flux density in Coulomb per meter square.
Let us take a uniform volume and place a lot of charges inside the volume V

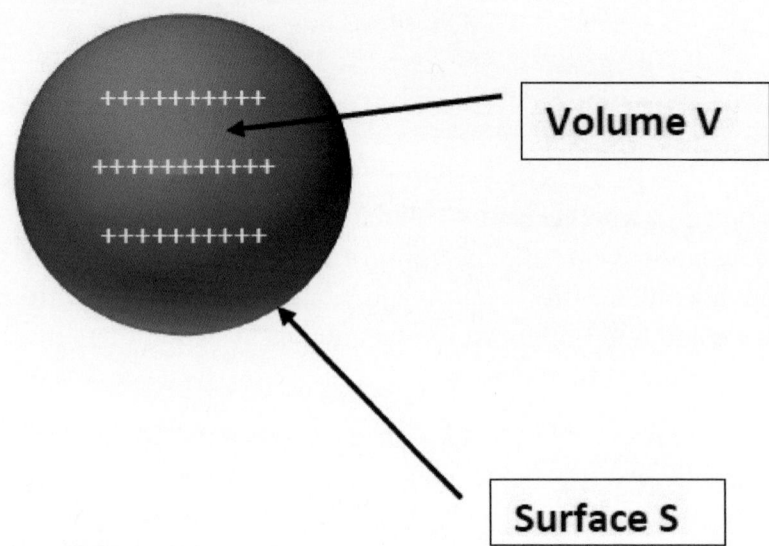

Let us define a new quantity called charge density inside the volume V

$$charge\ density = \frac{charge}{volume}$$

Since we have volume, the charge density becomes VOLUME CHARGE DENSITY

In equation terms it can be written as

$$\rho_v = \frac{Q}{V}$$

$$Q = \rho_v V$$

In the example above, we assumed **uniform** volume like sphere which contains charges. Other uniform shapes include box and cylinder

For **non-uniform** volumes, the volume cannot be obtained from a simple formula like sphere. We need to use calculus

We take small section of the volume and call it **dv** or differential volume. This differential volume contains small subset of the charges **dQ**

$$dQ = \rho_v dv$$

We can continue to divide the total volume into differential volumes and each will contain small subset of the total charges.

Total charge is obtained from integrating the differential volume charges

$$Q = \int \rho_v dv \qquad (2)$$

If you equate equation (1) and equation (2) we get

$$\int D.ds = \int \rho_v dv \qquad (3)$$

Equation (3) is called GAUSS'S LAW in electrostatics

In electromagnetics, the concepts of CURL and DIVERGENCE are very useful in calculations and writing Maxwell's equation in compact form.

If we have a vector field A such as electric field vector or magnetic field vector

Curl of A is written as $\nabla \times A$

Divergence of A is written as $\nabla . A$

The curl operation used vector cross-product while the divergence operation uses vector dot-product

Physical meaning of Curl is circulation of the vector field around a point. Example is circulation or rotation of magnetic field around a current carrying wire (Ampere's law)

Physical meaning of divergence is a vector field emerging and spreading from a point. Example is electric field lines diverging from point charges (Gauss law)

<u>**Two useful theorems from vector fields**</u>

If an object moves between two points along straight line of length l

Work done by the object = Force x distance

$W = F . l$

What if the distance moved is not straight line but a curve.

You could divide the curve into small straight line segments of length dl, multiple the differential distance by the force to get differential work done. Again take another differential distance dl multiply by force to get work and so on.

But you want total work done by the object right?

You can use calculus to find the total work along the total distance moved

$$W = \int F.dl$$

This type of integration is called line integration.

You could similarly perform surface and volume integration. These integration forms are summarized below

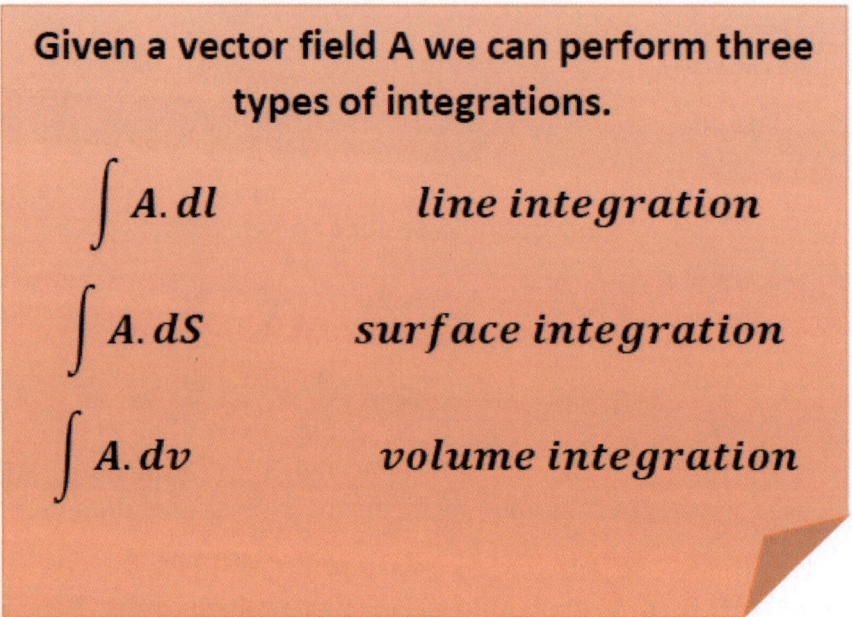

Given a vector field A we can perform three types of integrations.

$$\int A.dl \qquad \text{line integration}$$

$$\int A.dS \qquad \text{surface integration}$$

$$\int A.dv \qquad \text{volume integration}$$

These types of integration are related. This means line integration can be converted into surface integration and surface integration can be converted into volume integration

To do this integration conversion, Curl and Divergence will play a great role

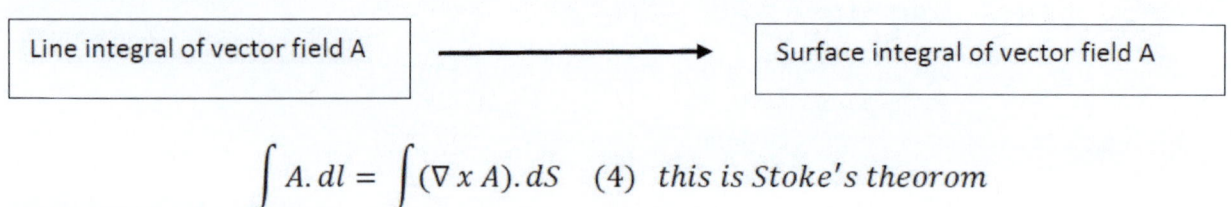

$$\int A.dl = \int (\nabla \times A).dS \quad (4) \quad \text{this is Stoke's theorom}$$

This means if I integrate a vector field A over a length (path) l it is the same as integrating the curl of A over surface S formed by the closed length (path).

Remember a closed path or length forms an open surface

Stokes' theorem will be useful for writing Ampere and Faraday's law in point form

| Surface integral of vector field A | → | Volume integral of vector field A |

$$\int A.dS = \int (\nabla . A)dv \quad (5) \quad \text{this is divergence theorom}$$

This means if I integrate a vector field A over a surface A it is the same as integrating the divergence of A over the volume enclosed by the surface.

Remember a closed surface forms a volume inside.

Divergence theorem will be useful in writing Gauss's law in electric and magnetic fields in point form.

Using equations (4) and (5) we can now write Gauss's law (3) in point form

Recall from (3) Gauss's law for electrostatics states that

$$\int D.ds = \int \rho_v dv$$

The LHS contains surface integral. Using equation (5) we can convert it into volume integral as

$$\int (\nabla . D)dv = \int \rho_v dv$$

For these two volume integrals to be equal, their integrands must be equal

Hence

$$\nabla . D = \rho_v \quad (6)$$

This is Gauss law in point form.

Equation (3) and equation (6) are the same. Both are Gauss's law. The first one is in integral form while the second equation is in point form

Static magnetic fields

Static magnetic fields are produced by currents. It was discovered that a wire carrying current produces magnetic field that will encircle the wire

Current → Magnetic field

I → H

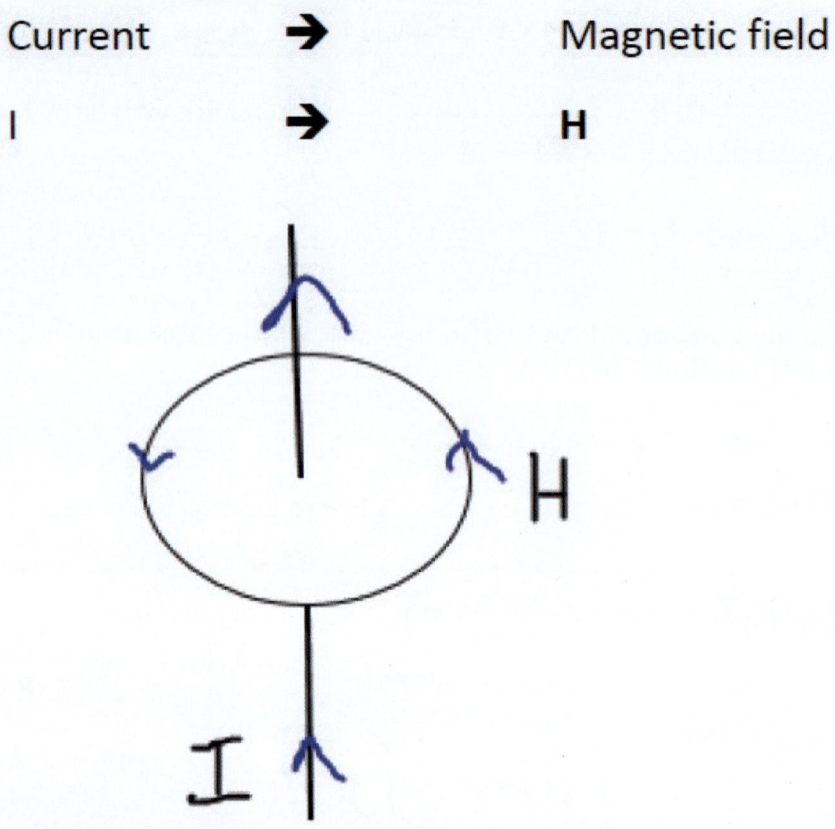

As you can see the magnetic field follows along a closed path around the wire. The current enters and exits the path of the magnetic field formed.

Since the magnet field H is measured in ampere per meter (A/m) from Biot-Savart law, you could say that

$$I = H.l$$

Where l is the length of the loop and it equal the circumference of the circle

Hence

$$I = H.2\pi r \qquad r = radius\ of\ the\ circle$$

The path above is perfect circle. However in reality we are interested in more general and irregular path like the one below called Ampere path

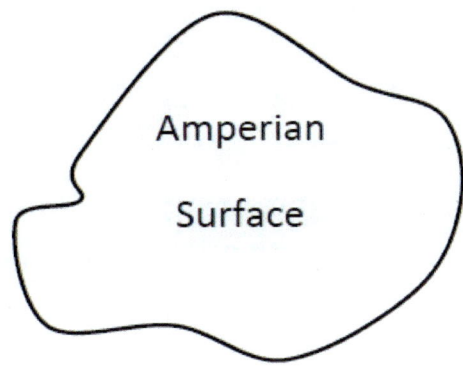

Divide this path into differential lengths each of straight line approximation of length dl

Multiple each differential length dl by H

And add them together using integral Calculus

$$I = \int H.dl \qquad \text{Ampere's law} \quad (7)$$

N.B H is vector quantity and is given right hand rule. Also closed path l is a vector, and its direction at each point of the direction of the vector tangent at that point.

Current flows along wires and conductors. In theory a wire is drawn as a line of length l and zero thickness, hence zero surface area.

But practically a wire has a thickness and it has a surface area S. Current I follows along the conductor surface.

It is good to know how much current passes along small subset of the area, and the quantity that will help us determine this is surface current density J

$$surface\ current\ density = \frac{current}{area}$$

$$J = \frac{I}{S} \qquad in \qquad \frac{A}{m^2}$$

$$I = J.S$$

This expression is enough for simple surfaces such as cylinder surface, sheets and so on.

But in space we normally have an arbitrary surface of any shape that is irregular.

So we divide the surface into differential surfaces dS.

Each differential surface dS carries small subset of the total current

$$dI = J.dS$$

Total current is obtained from integrating this equation over the entire surface

$$I = \int J.dS$$

Substitute this into equation (7)

$$\int J.dS = \int H.dl \qquad Ampere's\ law$$

N.B the surface S has an orientation is space. Hence it is vector quantity.
Do you remember Stokes' theorem?

Well it converts line integral of a vector field into surface integral of the curl of the vector field

$$\int H.dl = \int (\nabla \times H).dS$$

Using this, Ampere's law in (7) now becomes

$$\int J.dS = \int (\nabla \times H).dS$$

For these two integrals to be equal, the integrands must be equal

Hence

$$\nabla \times H = J \qquad (8)$$

Equation (8) is point form of Ampere's law and it is the third equation of Maxwell equations

So far we have defined electric flux density D and used it to derive Gauss's law. Now let us defined magnetic flux density B.

An external bar magnet produces magnetic field which starts at North Pole and ends at South Pole as shown below

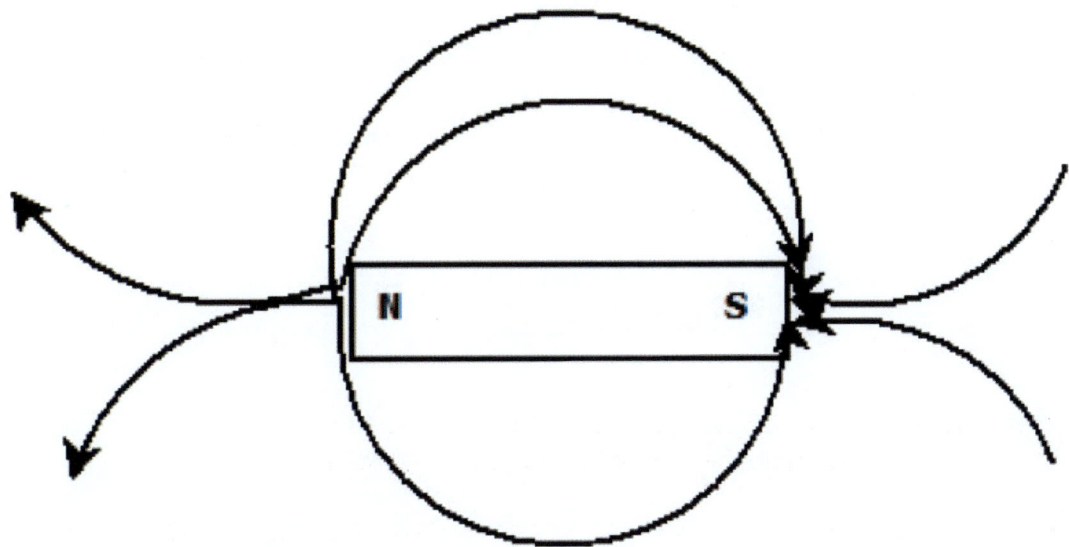

Let us place a close path (open surface) in front of the magnet and allow the fields to pass through the surface S as shown below

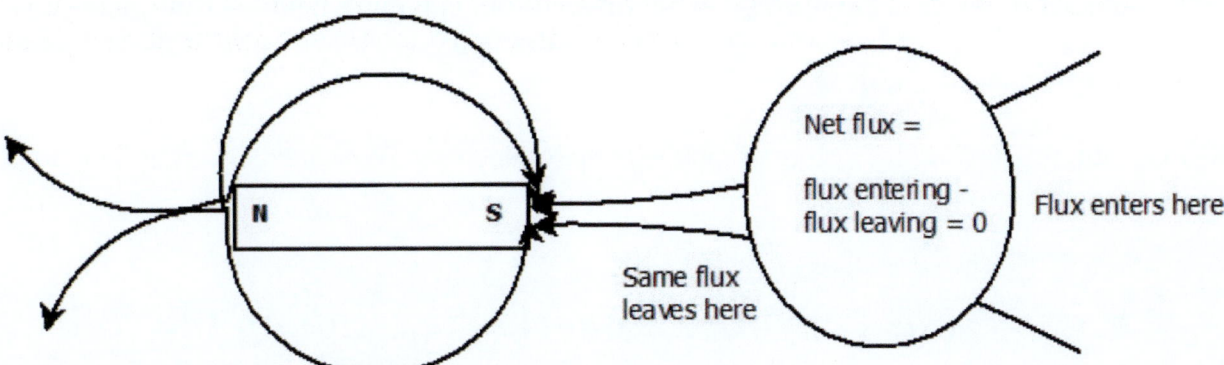

As can be seen above, the magnetic flux enters and exits the surface. There is so source inside that generates the flux. The flux comes from external magnet. This suggest that magnetic flux

through a surface is zero. You can drive the same conclusion if you replace the bar magnet with current carrying wire

Net magnetic flux passing through a surface is ZERO.

$$\int B.dS = 0 \quad (9)$$

This is Gauss's law in static magnetic fields.

To write this into point form, we need to replace the surface integral with volume integral using divergence theorem

$$\int B.dS = \int (\nabla . B) dv = 0$$

The only way the volume integral be zero is if the integrand ($\nabla . B$) is zero

$$\nabla . B = 0 \quad (10)$$

This is the second Maxwell equation.

The fourth Maxwell equation is based on Faraday's law. Faradays law can be stated as

When a magnetic field changing with time passes through a coil of wire, an EMF (voltage) will be induced in the coil. For voltage to be induced, the magnetic field must be changing (moving towards or away from the coil). The opposite will give same results if the magnet is stationary and the coil is moved.

The magnetic flux linking the coil is a function of space and time. We can take partial derivative of B with respect to time

$$\frac{\partial}{\partial t} \int B.dS = time\ varying\ magnetic\ flux\ linking\ the\ coil$$

The voltage induced in the coil by the external magnetic field will produce and electric field. Time varying magnetic field ➔ voltage in coil ➔ electric field in coil.

Electric field E and voltage V are related by the length of the coil loop in which the current flows

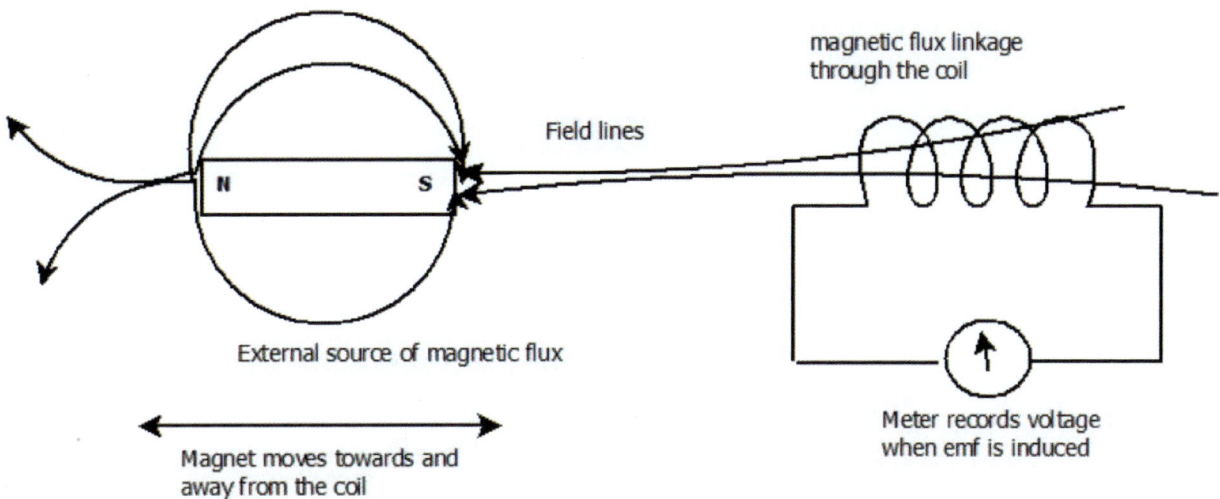

$$V = \int E \cdot dl$$

Since the changing magnetic field produced the voltage, the above two equations are equal

$$\int E \cdot dl = -\int \frac{\partial B}{\partial t} \cdot dS \quad (11)$$

Faraday's law

The minus sign is from Lenz's law

We can write this equation in point form using Stokes' theorem

$$\int E.dl = \int (\nabla \times E).dS$$

$$\int (\nabla \times E).dS = -\int \frac{\partial B}{\partial t}.dS$$

Hence

$$\nabla \times E = -\frac{\partial B}{\partial t} \quad (12)$$

Faraday's law in point form

The four electromagnetic quantities E, D, B, and H are related by permittivity and permeability

$$D = \varepsilon E$$

$$B = \mu H$$

A summary of electromagnetic equations that govern electromagnetic waves are presented below

Fundamentals of electromagnetic waves theory

Maxwell equation in point form	Maxwell equation in integral form	Law it based on
$\nabla \cdot D = \rho_v$	$\int D \cdot dS = \int \rho_v dv$	Gauss's law of electrostatics: Net electric flux passing through as closed surface is equal to net charge enclosed in the surface. D = electric flux density in $\frac{C}{m^2}$ ρ_v = volume charge density in $\frac{C}{m^3}$
$\nabla \cdot B = 0$	$\int B \cdot dS = 0$	Gauss's law in magnetostatics: Net magnetic flux crossing a given surface is zero. This mean there are no magnetic monopoles (point charges) B = magnetic flux density in Tesla
$\nabla \times H = J + \frac{\partial D}{\partial t}$	$\int H \cdot dl = \int (J + \frac{\partial D}{\partial t}) \cdot dS$	Ampere's law: Current carrying wire produces magnetic field around the wire. Magnetic field around the path enclosing the surface is the same current density distributed over the surface H = magnetic field intensity in $\frac{A}{m}$ J = surface current density in $\frac{A}{m^2}$
$\nabla \times E = -\frac{\partial B}{\partial t}$	$\int E \cdot dl = -\frac{\partial}{\partial t} \int B \cdot dS$	Faraday's law: A time varying magnetic field linking a coil of wire induces and electric field in the wire. E = electric field intensity in $\frac{V}{m}$

These four laws govern the behavior of electromagnetic waves such as those of microwave transmission signals. An electromagnetic wave is composed of fluctuating electric and magnetic fields that are perpendicular to the direction of wave travel

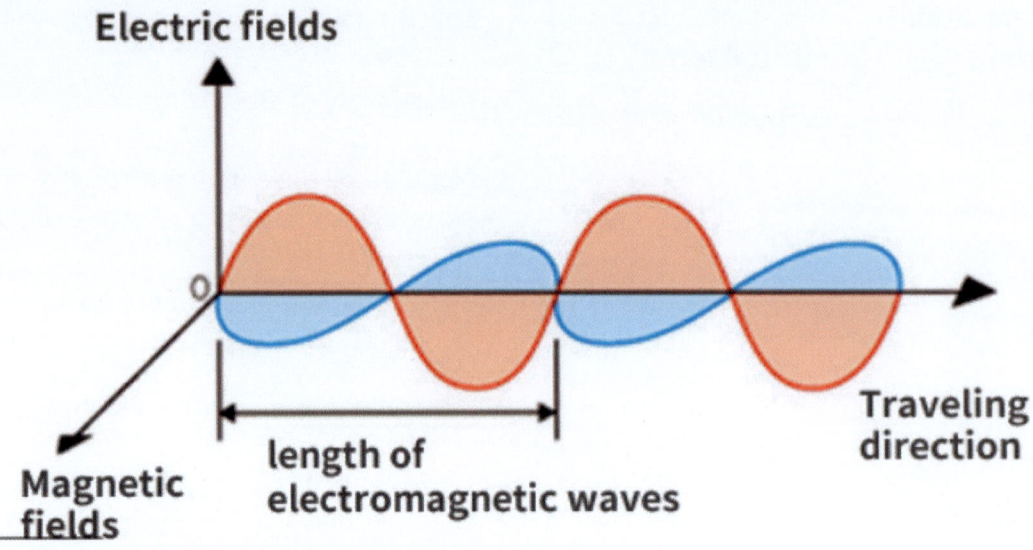

Figure 1.1 an electromagnetic wave

There are different electromagnetic waves each characterized by frequency, wavelength and phase. Such we don't just have one electromagnetic wave rather we have electromagnetic spectrum

Different names are given to different waves in the electromagnetic spectrum depending on the frequency and wavelength. Each portion of the spectrum has got variety of applications such as cellular, satellite, conventional microwave, millimeter microwave, optical fiber and radar

The frequency ranges of the radio electromagnetic spectrum 3 kHz – 300GHz is given below

Table 1.1 electromagnetic spectrum for radio communication

Designation	Meaning	Range	Application
VLF	Very low frequency	3 – 30kHz	submarines
LF	Low frequency	30 – 300kHz	AC line
MF	Medium frequency	300kHz – 3MHz	AM radio
HF	High frequency	3 – 30MHz	Weather, aviation,
VHF	Very high frequency	30 – 300MHz	FM radio

UFH	Ultra-high frequency	300MHz – 3GHz	TV, cellular, radar
SHF	Super high frequency	3 – 30GHz	Satellite, microwave, radar
EHF	Extremely high frequency	30 – 300GHz	Microwave

Digital communication link performance

In this books our focus will be mainly on digital communication when we will discuss microwave, Satellite and cellular system. Digital communication uses binary to transmit data. While in analog communication we worked on waves, in digital communication we will work on symbols

In analog communication we normally used signal power over noise power ration (S / N)

In digital communication, we use bit energy per noise spectral density (E_b/N_o). This performance metric can be related to S / N by the following equation

$$\frac{E_b}{N_o} = \frac{S}{N} \times \frac{W}{R}$$

Where

   S = signal power
   N = noise power
   W = bandwidth
   R = data rate
   E_b = bit energy
   N_o = noise spectral density

To measure the performance of the digital communication link, the error probability is plotted against E_b/N_o

Fundamentals of electromagnetic waves theory

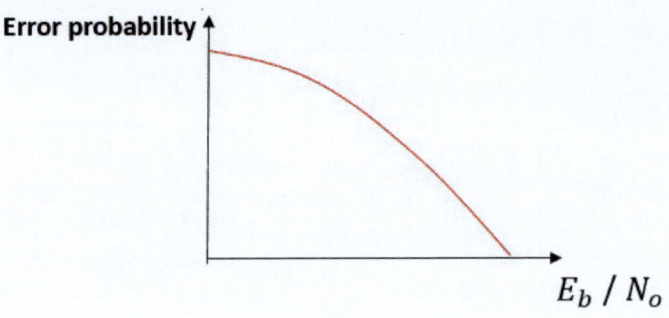

Link margin = received $\frac{E_b}{N_o}$ **- target** $\frac{E_b}{N_o}$

From here we can drive the link margin equation based on the following link parameters

```
EIRP
Receive antenna gain = Gr
Data rate = R
Target E_b/N_o
Path loss = L
Other losses = L_o
Noise temperature = T_o = 290K
Boltzmann constant = k = -228.6
```

Link margin = EIRP + Gr − Target $\frac{E_b}{N_o}$ **- kT_o + R − L -** L_o

The design engineer can then trade-off parameters in the above equation based on design objectives

For example

- Reduce EIRP to get rid of excess link margin
- Increase data rate by reducing Target $\frac{E_b}{N_o}$ (by manipulating ModCod)

Chapter Two

Introduction to Microwave transmission

Telecommunication network can divided according to geographical area in which it can provide services. There are local area networks (LAN) that cover in limited geographical area such as enterprise building or home network. Since the network components in LAN are placed in close proximity, short Ethernet cables connect the devices. In this case, the Ethernet cable acts as Transmission network.

When we have networks that are geographically dispersed, some means of connection must be designed, or these networks will be isolated from each other.

Transmission links are everywhere in Public and private networks. Microwave networks are used in many cases for its simplicity in design and deployment compared with Fiber.

Examples of where microwave transmission systems are applied include

- Headquarter of business company in city A is connected to branch office in city B, where cities A and B are 100km apart.
- In cellular systems, large number of base stations are placed in different areas of the country to provide mobile voice and internet coverage. These stations must be connected to the core network that maybe in the same area or distance away.
- A large ISP may provide WAN links to government sectors to interconnect different departments of the government or connect them to the internet.
- Two large mobile operators in two neighboring countries may install long chain of microwave links for roaming.

Some of microwave transmission applications are depicted in the following diagrams

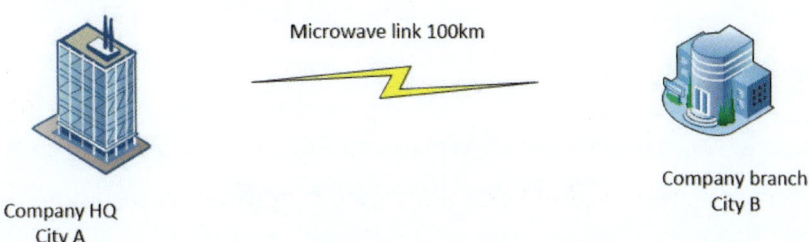

Figure 2.1 Microwave link connecting company HQ and branch separated at 100km

Introduction to Microwave transmission

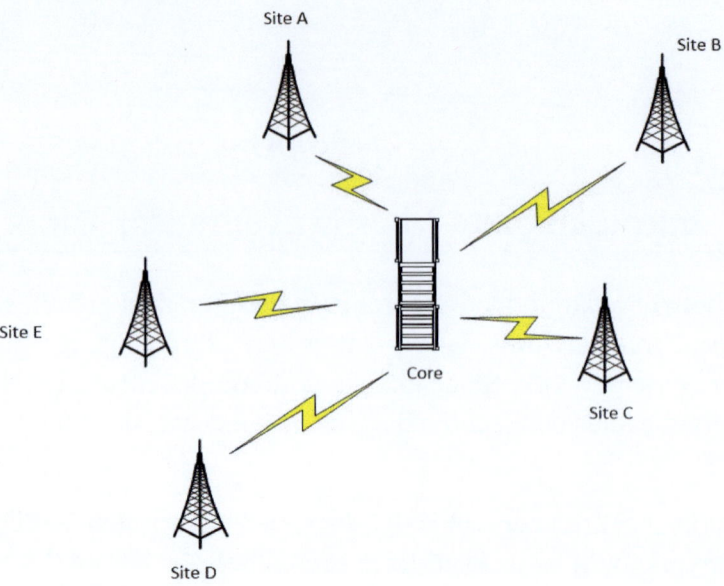

Figure 2.2 Microwave links connecting mobile base stations to core network

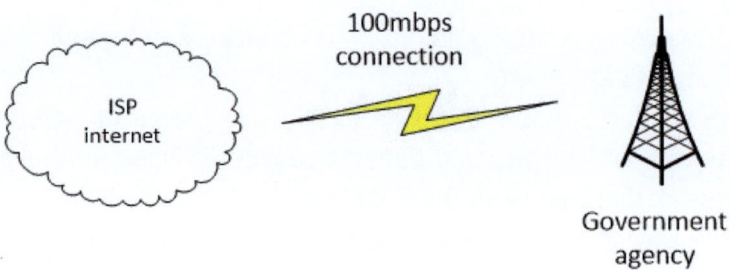

Figure 2.3 Private microwave links

Just as everything need resource to work, microwave links need bandwidth to transport data from one location to another. The amount of bandwidth allocated will determine the capacity of the link. More on this will be covered in frequency planning chapter.

Overview of microwave link design

In this section we take a look at various steps taken to design a microwave link. A basic microwave consists of a transmitter and receiver with wireless as medium as shown below.

Introduction to Microwave transmission

Figure 2.4 basic microwave link

Today's microwave links are digital and are full duplex. Each link carries both transmit and receive signal simultaneously. The fundamental design goal of any microwave link is to calculate the receive signal power we get at one site when transmit power is sent from another site.

The transmit power is reduced (attenuated) by various phenomena associated with the wireless systems and include:

- Wireless channel loss (free space pathloss)
- Terrain and man-made obstacles (obstruction fading)
- Ground reflections (multipath fading)
- Ducts (atmospheric multipath)
- Atmospheric gaseous absorptions
- System and equipment losses
- Interference loss

Total loss is the summation of all the losses mentioned above and cause transmit signal to be reduced before it is received.

The equation that links transmit power, total loss, and receive power is called link budget.

The link budget equation is stated as

Receive power = Transmit power – Total loss (RX = TX – L).

Since RX power can be small, it is good point of exercise to define the smallest RX power below which reception of data is impossible. The smallest RX signal for which the radio can accept to process data is called receiver sensitivity. For this reason all microwave radios have RX sensitivity parameter.

For a communication link to work properly the RX signal must be well above Rx sensitivity. The amount of power by which Rx signal is greater than RX sensitivity is called link margin and must be adequately enough. Large link margin will protect a link against losses.

Rx power = Rx sensitivity + link margin.

A microwave link also consists of radio equipment and an antenna. The radio unit could be placed indoor (split-mount) or outdoor (all outdoor). Specification required for the microwave radio and antenna are listed below

Radio equipment specifications	Antenna specifications
TX power	Gain
Rx sensitivity	Frequency band
Modulation supported	Beamwidth
Channel bandwidth	Diameter
Data rate	Polarization
User interfaces	XPD
	F/B

So when transmitting a signal, the radio TX power is amplified by the antenna gain. Thus, what is actually transmitted is radio TX power plus antenna gain minus feeder loss and it is called effective isotropic radiated power (EIRP)

EIRP = TX power + transmit antenna gain – feeder loss

Link budget = Rx = EIRP – Total loss

On the receive side, RX power is added to receive antenna gain.

The final link budget equation is

RX power = EIRP – Total loss + RX antenna gain

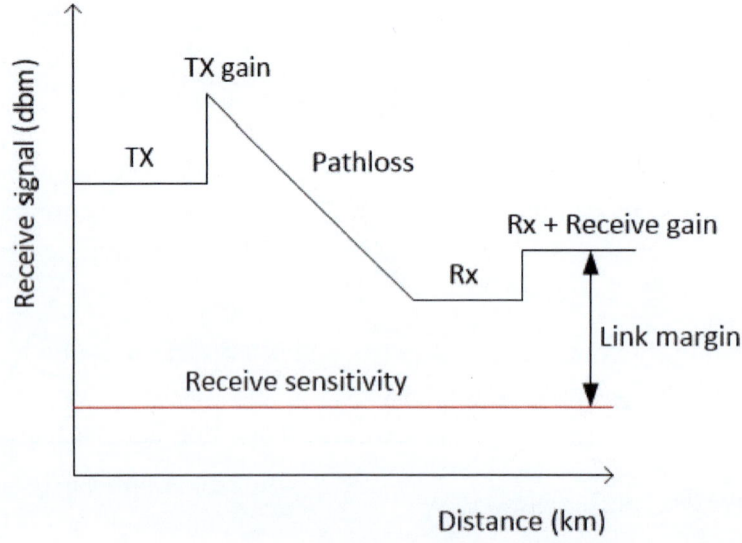

From this graph, RX signal level should be above receive sensitivity by an amount called link margin. A good microwave link will have sufficient link margin.

Introduction to Microwave transmission

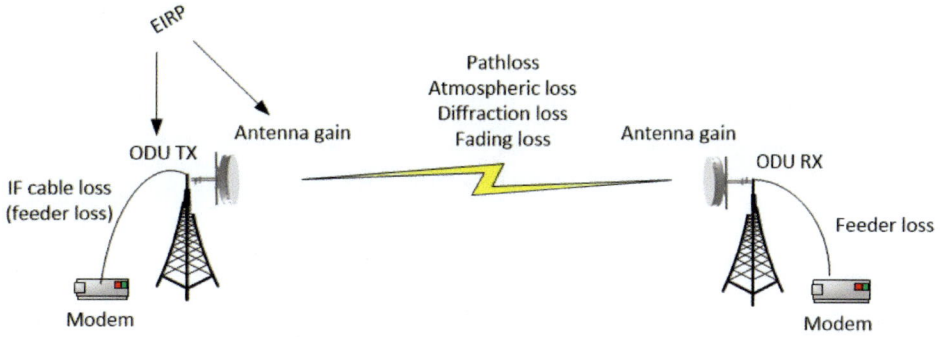

Link budget = RX = EIRP – Total loss + RX antenna gain

Now we will define the antenna parameters mentioned in table 1. These antenna parameters are documented in vendor datasheet.

- **Gain** of an antenna is a performance related parameter of the antenna. It is the ability of the antenna to radiate power in a particular direction. The gain value of an antenna determines how well the antenna converts input energy from transmitter into an electromagnetic wave. It is measured in dBi

  A given antenna can either radiate power in all directions (Omni directional antenna) or just in a particular direction (directional). Microwave antennas are directional point to point.

Omni directional antennas
Radiate in all directions.
Used in land mobile systems

Directional antenna focus energy in a particular directions. Used in Microwave links

- **Frequency band** is the range of frequencies the antenna can transmit and receive. For example an antenna may work in 7.4 – 7.9 (7 GHz) range or 14.4 – 15.35 (15 GHz) range.
- **Diameter** of antenna is the size of the antenna. It can be 0.3m, 0.6m, 0.9m, 1.2m, 1.8m, 2.4m, 3m

  Antenna gain, frequency band, and diameter are related. To get higher antenna gain, increase the diameter and frequency band.

- **Polarization** is the orientation of electric field vector component of the microwave signal. In microwave system, the signal can be transmitted or receive in either vertical polarization or horizontal polarization.
- **XPD** is cross polar discrimination value of the antenna. Dual polarization (both vertical and horizontal) antennas can transmit and receive two cross polarized signal at the same frequency. It is possible to transmit and receive same radio channel in both polarization. XPD is the ability of the antenna to discriminate the two polarizations. An antenna should have high XPD value to perform well. Its unit is dB and typical value is 30dB
- **F/B** is front to back ratio of an antenna. It is the ratio of power radiated in main lobe to that in back lobe. It is measured in dB. A typical value for high performance antenna is 75dB.

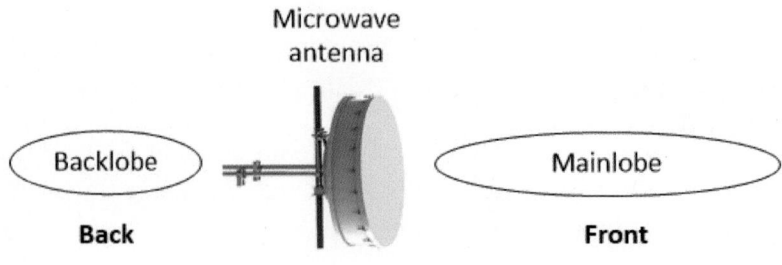

Front / back = F / B ratio

Next we define the radio parameters mentioned in table 1.

Please bear in mind by radio we mean microwave outdoor unit (ODU) which does the functions of transmitter/receiver and power amplifier. These radio parameter are documented in the vendor equipment datasheet.

- **TX power** is RF power generated by the ODU when low frequency (IF) signal is input from indoor unit (modem). It is measured in dBm and typical value is 24
- **RX sensitivity** is the minimum signal power a radio receive can accept to process data. It is measured in negative dBm and typical value is -80
- **Modulation order supported** can be 64-QAM for medium capacity links all the way to 4096-QAM for high capacity radio trunks that carry a lot of data.
  Modulation is important in radio links and it is the process in which a user signal is carried by a high frequency signal generated by an electronic oscillator circuit. The user data varies the amplitude, frequency or phase of the carrier. Hence amplitude modulation, frequency modulation and phase modulation.
  QAM is quadrature amplitude modulation and varies both amplitude and phase of the carrier signal.

Modulation order supported by the radio		
16-QAM	4 bits of data carried inside 1Hz of bandwidth	
32-QAM	5 bits of data carried inside 1Hz of bandwidth	
64-QAM	6 bits of data carried inside 1Hz of bandwidth	
128-QAM	7 bits of data carried inside 1Hz of bandwidth	
256-QAM	8 bits of data carried inside 1Hz of bandwidth	As modulation order increases, data rate increases.
512-QAM	9 bits of data carried inside 1Hz of bandwidth	
1024-QAM	10 bits of data carried inside 1Hz of bandwidth	
2048-QAM	11 bits of data carried inside 1Hz of bandwidth	
4096-QAM	12 bits of data carried inside 1Hz of bandwidth	

- **Channel bandwidth** is the capacity of the radio channel. The wider the radio channel, more information can be carried. Higher modulation order and wider channel capacity will guarantee higher data rate.

  When planning links, it is important to decide link capacity requirement as that will determine what channel bandwidth the link will need.

- **Data rate** will result from a combination of channel bandwidth and modulation order. A typical data rate of microwave link for LTE backhaul could be 500Mbps. To achieve this, get radio equipment that will support 256-QAM at 56 MHz bandwidth for example. Ultimately the vendor will give the operator support on this as the former has deeper expertise than the latter.

There is tradeoff between capacity needed for new link and transmission distance of the link hop. Nowadays the world is moving to a high capacity networks such as LTE and 5G. So techniques used in the past for links greater than 50km may not work in current requirements. Higher transmission capacity typically employ higher order QAM. Higher QAM is subject to noise and rain attenuation and will be suitable in shorter links. Many microwave radio vendors offer high powered ODUs that can increase transmission distance. Other techniques that also increase distance include increasing antenna size and gain. These enhancement come with an increase in CAPEX (capital expenditure).

We wrap up this chapter with a typical example showing how transmission department functions are structured to manage effectively is shown below

Introduction to Microwave transmission

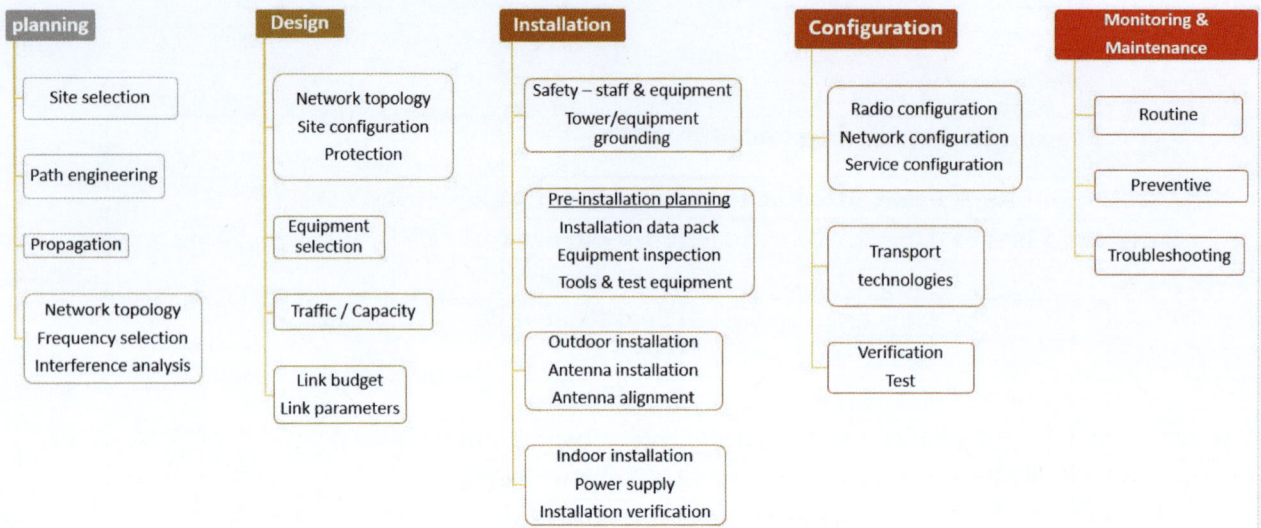

Chapter Three
Project management

Planning stage is the first step taken after inception of rolling out major transmission and RF network by new telecom operators. It is also used in expansion phases of already established operator.

In project management for new operator, things done include

- Conceptual framework of the overall project is laid out
- Resources needed are identified
- Project schedule is developed with cost estimation
- Market data collection
- Pre-marketing and exposing the operator name
- Hiring technical and operational staff
- Procuring network planning tools

Often new operators face challenge of balancing time and initial capital. As project timeline extends further beyond unreasonable schedule, cash flow increases and capital shrinks which exposes the new operators both operational and investment risks.

For mobile network operator, the part that cause significant delay is the preparation of RF and transmission infrastructure. Late planning of the network also delay purchase and delivery of equipment.

So hiring RF and transmission planner should be done immediately after operator decides launching telecom business. As preparation of sites starts, also acquisition of spectrum for licensing is carried out. If regulators authority data is not available which is not in most cases, then spectrum monitoring should be carried out by the operator itself which increases project cost.

The RF and transmission planner outlines the following project plan

- Receives requirement from the management (number of sites needed in first phase, spectrum licenses for both RF and transmission, and services the network is planned to support in the end)
- Assess and studies different ways of realizing the requirement with their associated pros and cons
- Orders and acquires both RF and microwave planning software and hardware tools (include Pathloss, RF tool, compass, GPS unit, multimeter, ground resistance meter, tape and digital camera)
- Usually the RF planner and the transmission planner are separate and done by different engineers.
- The RF planner first provides location of sites. Then the transmission planner will use the sites to determine required tower height

- The RF planner will now start coverage planning for sites and produce RF plan (capacity and spectrum plan, BTS configurations, hardware and software licenses, quality and performance data, cell RF template sheets)
- The transmission planner will start transmission links plan based on (capacity requirement from the RF plan, number of ODU and antennas for each link, link configuration, hardware and software licenses, future expansion plan, and link parameters)
- The technical manager will constantly get input from the RF and transmission planner and finally negotiate vendors following recommendations of the planning team)

Now the RF and transmission planning phase is tedious and demands high level of experience. Wrongs made in the phase will subsequently increase project implementation cost and affect long term objectives of the operator

What the planners do is to first list down all the requirement and tasks and then produce project schedule to make timing and coordination effective.

As sample of transmission planner daily project schedule is shown below prepared in MS project

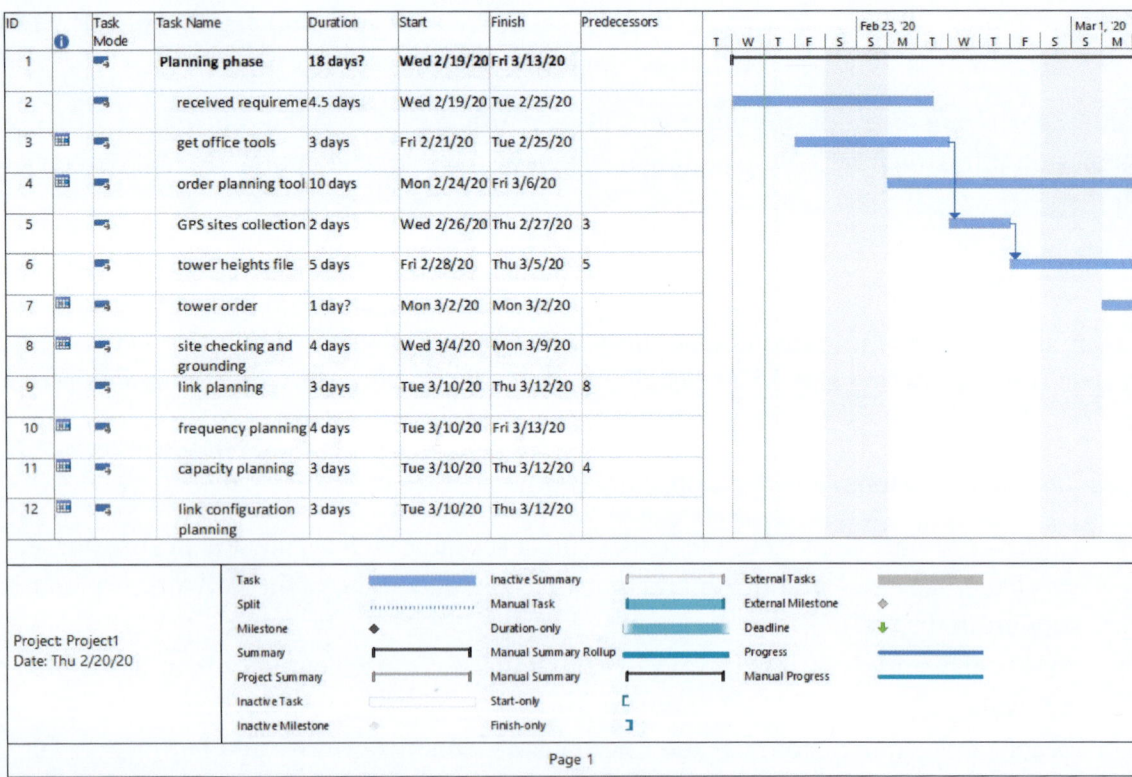

Figure 3.1 Microwave planning schedule

The resources needed in the planning phase is illustrated below. These resources should be acquired immediately after operator starts business.

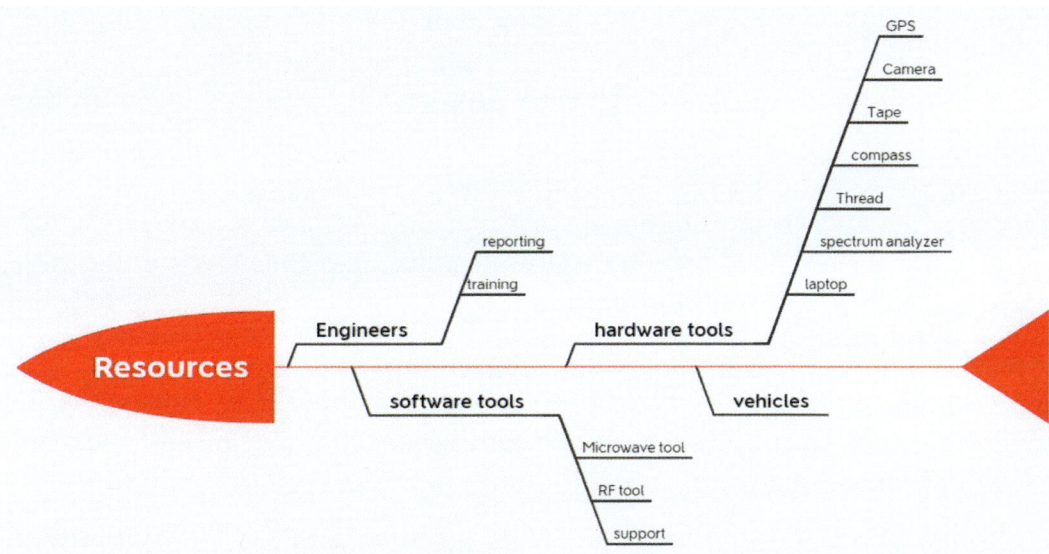

Figure 3.2 microwave transmission project resources

Factors that lead to project delay include

- lack of proper definition of project requirement
- Project scope is not clearly defined. Everything is simplified. Management did not break down the project into minor details. Many overseen delays will appear in the latter stages of the project
- Hiring all-inexperienced personnel for the sake of lower wages. Companies who tried this approach incurred more than their initial investment plan. It is always good to hire experienced people
- In the early phase of the project, cost benefit analysis not conducted, not analyzing project contents and attributes, not developing proper information management system, and not stating clear project objective all lead to failure or project cost beyond damaging threshold
- Project activities and proper allocation of resources to those activities is not monitored and controlled
- Degree of uncertainty of proposed plan, monitoring and evaluation gap, bad decision makings, lack of coordination and communication also can lead to project delay
- One of the most important things which delay project is lack of transparency and accountability. It also includes when the company witnesses micro-management and staff lose confidence which could lead to disarray of activities and team work

Chapter Four

Route design

In the last chapter, we have discussed the overall framework of microwave link planning, design, and implementation in the context of project management. We have seen when the microwave project is properly planned and all the activities scheduled, implementation of the project will become less difficult.

In this chapter we discuss how to plan route of the microwave network. In cellular systems, core switching systems are placed in a city, and another one in a remote city. Microwave links will then connect these two cities that are geographically separated. Route planning is required to guarantee LOS (Line of sight).

Clearance planning of the hops above terrain is the main critical step in route design. Route clearance will then determine tower height required.

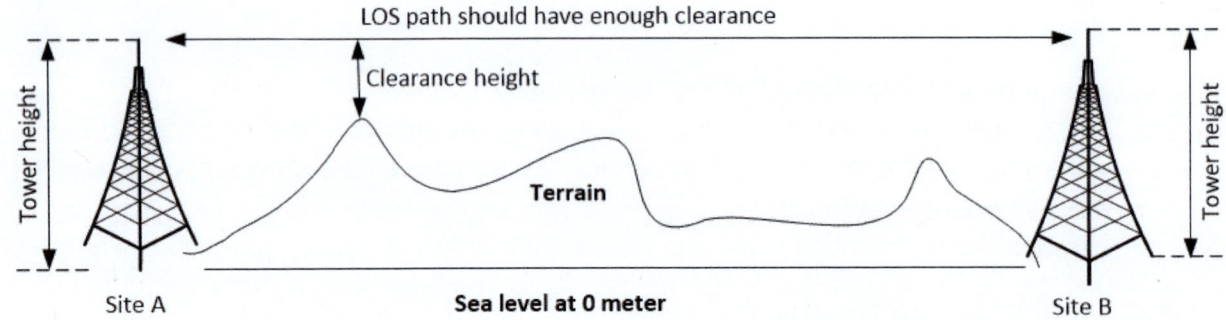

Figure 4.1 path clearance for micrwove LOS

One of the first steps in route design is site survey. A team of transmission engineers will go out and visually study the terrain. Four categories of terrain could be identified as

- Water body.
- Flat land
- Gently raising land.
- Mountainous and hilly terrain.

Each type of terrain exhibit different propagation of the microwave signal.

Table 4.1 terrain types for microwave propagation

Terrain	Propagation effects
Water body	Signal reflection, multipath fading. Poor for microwave propagation
Flat land	Signal reflection, multipath fading. Poor for

	microwave propagation
Gently raising land	Signal reflection can occur for long links. Better than the previous two
Hilly terrain	This terrain is good for microwave propagation. Signal reflection can be overcome by taking advantage of the mountains. The disadvantage is site accessibility may become difficult if there are no roads in the mountain.

Many operators take the main road connecting cities as a route. The result will be linear route where links are daisy chained.

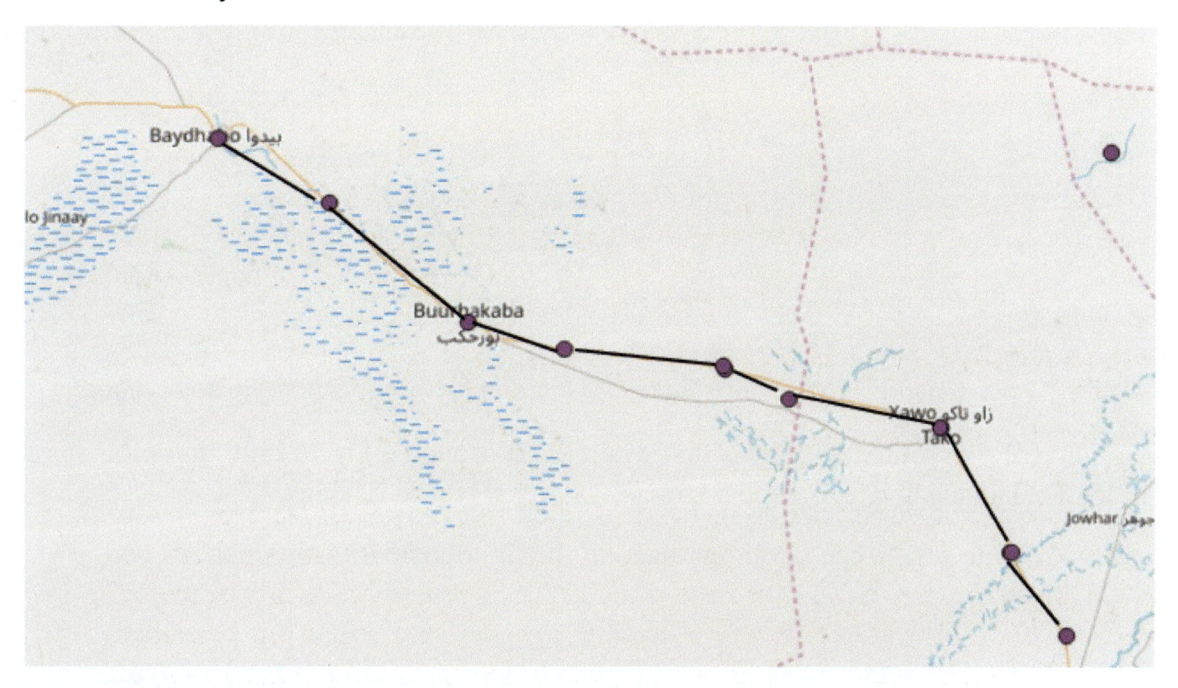

Figure 4.2 single route for backbone connectivity between Mogadishu and Biadoe

A new route can be dropped from the main road if there are other cities that need service.

Another type of route is ring route used whenever redundancy paths are needed.

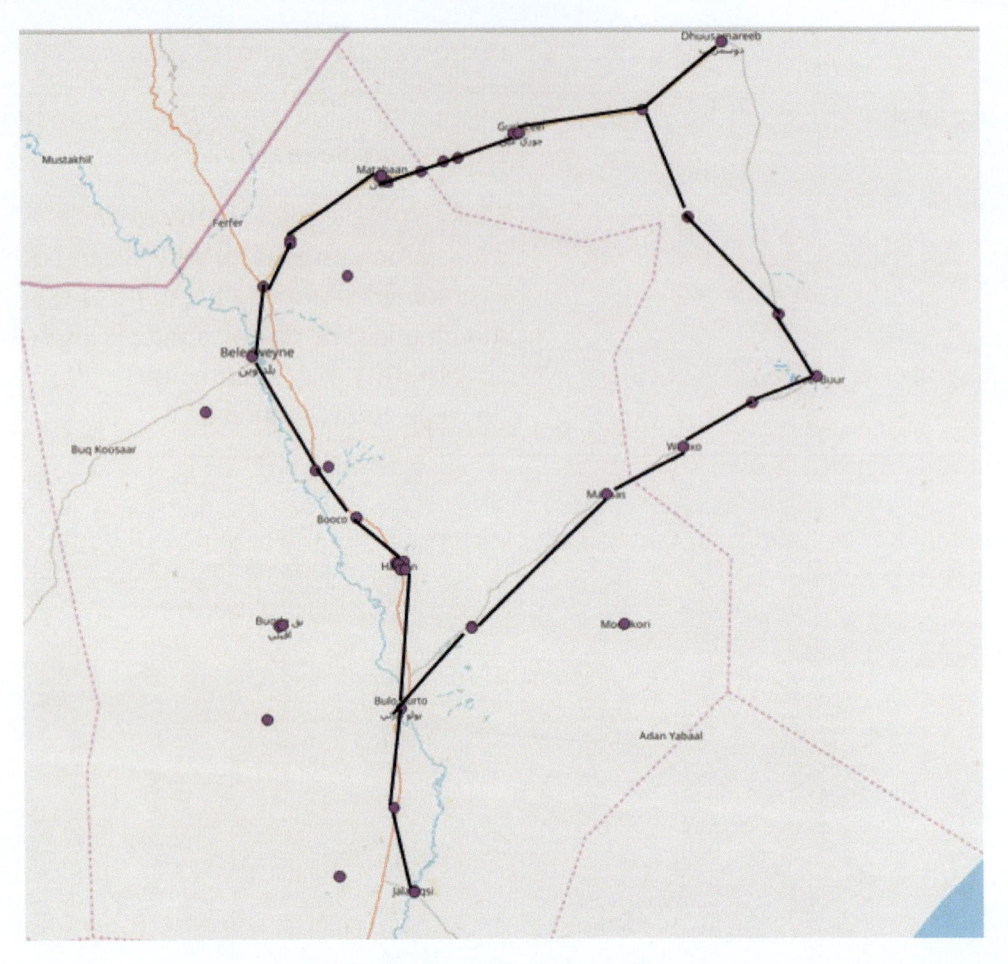

Figure 4.3 two redundant backbone routes forming ring between Mogadishu and Galgadud

The advantage of ring route topology is that when IP links are deployed, routing protocols can perform load balancing and automatic traffic re-routing if one route fails.

The process of route design can be summarized in the figure below

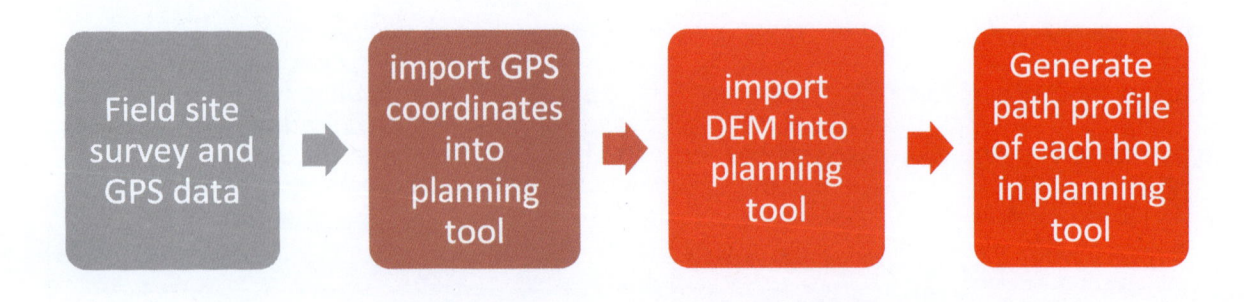

Field site survey and GPS data → import GPS coordinates into planning tool → import DEM into planning tool → Generate path profile of each hop in planning tool

Figure 4.4 microwave route planning process

Route design

The process starts when a team of transmission engineers go out in the field after route planning activity has been approved by the project manager. Usually the starting and the ending points of the route is made clear before survey. The route normally starts at the operator headquarter and ends in branch offices of the operator in remote cities.

GPS coordinates of any potential site is taken. One such GPS unit the author has used over the years is shown below

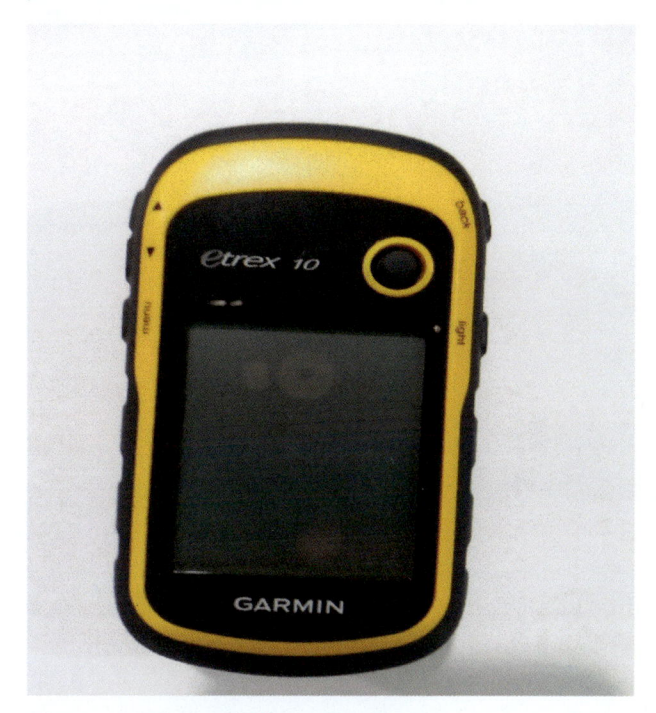

Figure 4.5 GPS unit for surveying

The coordinates (latitude/longitude) data of the sites is then reported as shown below.

Table 4.2 latitude and longitude of site location

Site name	Latitude	Longitude
Mogadishu	02 05 05.76 N	045 14 31.32 E
Balcad	02 21 26.37 N	045 23 03.31 E
Jowhar	02 46 56.20 N	045 29 47.89 E

The sites information is now imported into planning tool. There are many good planning tools in the market. The author has used Pathloss tool over the years and will thus be used in the illustration. When site coordinates are input into Pathloss, it will look like this

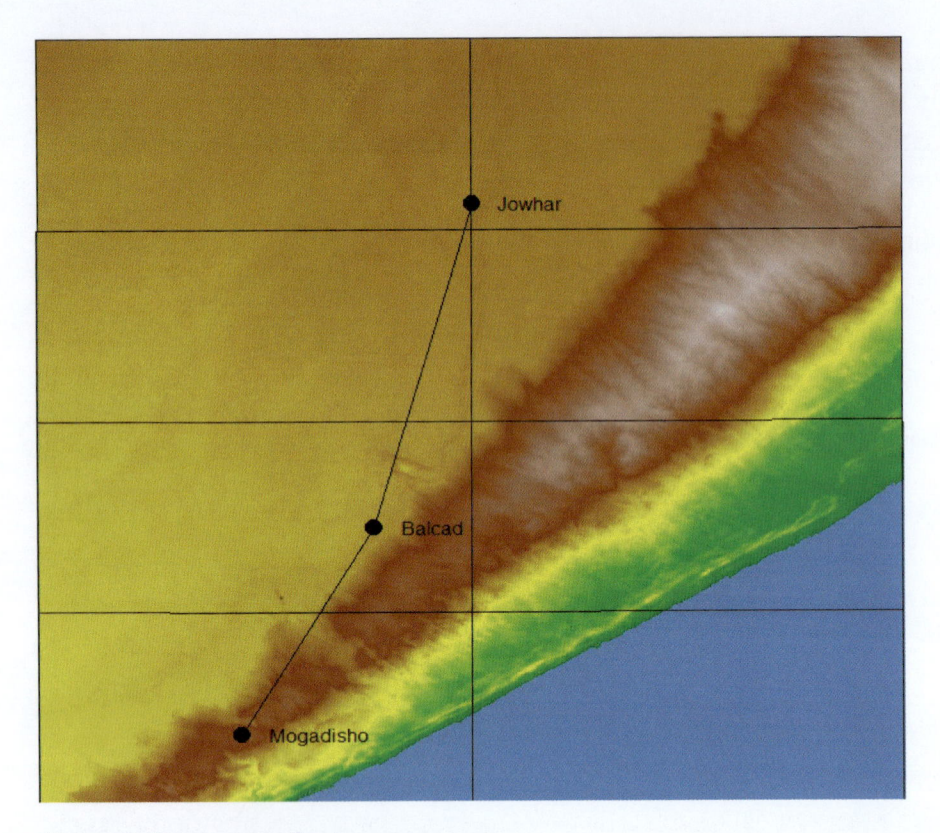

Figure 4.6 example of microwave route

The planning tool will not function properly until digital elevation database is imported. The elevation data will be used by the tool to project and geocode sites. It contains elevation heights at each point on earth surface.

In this project we will use the most prominent global DEM called SRTM. Shuttle radar topography mission (SRTM) is a high resolution digital elevation data with (.hgt) file extension, and can be freely downloaded from internet. The 1 arch version is used.

Route design

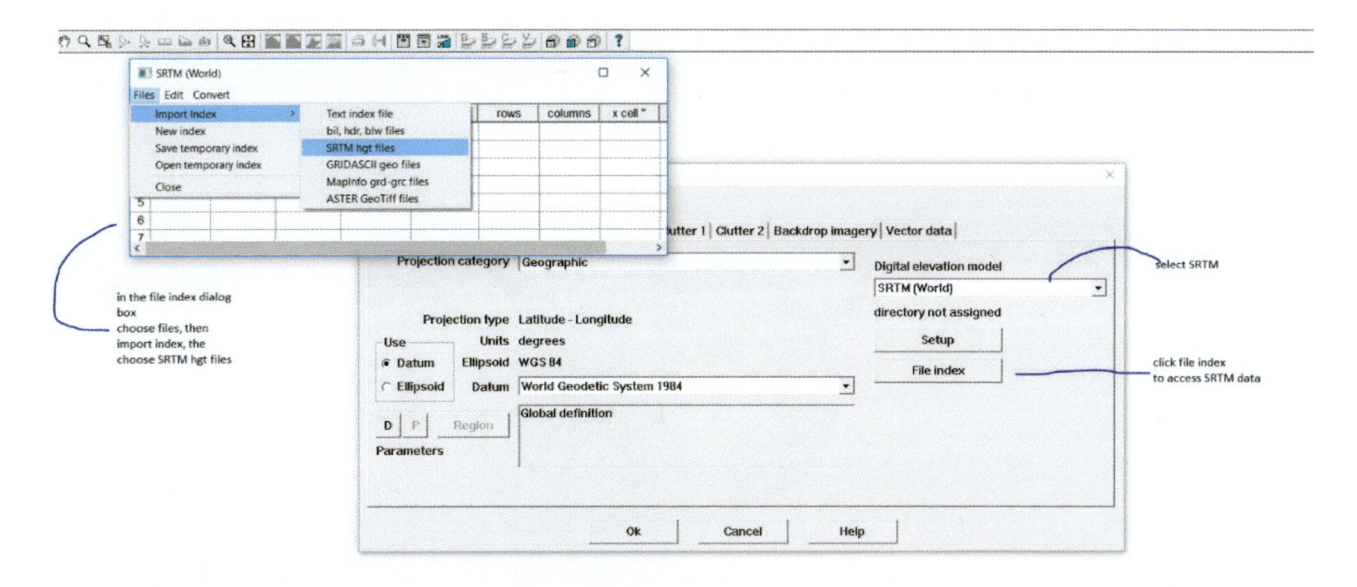

Figure 4.7 using SRTM GIS for microwave route planning

The most recent SRTM data is with resolution of 30m.

Now the sites are imported into planning tool and SRTM elevation set as primary terrain database, we can check LOS between sites and get required tower heights.

Terrain profile generated by Pathloss 5.1 for sites Mogadishu and Balcad is shown below

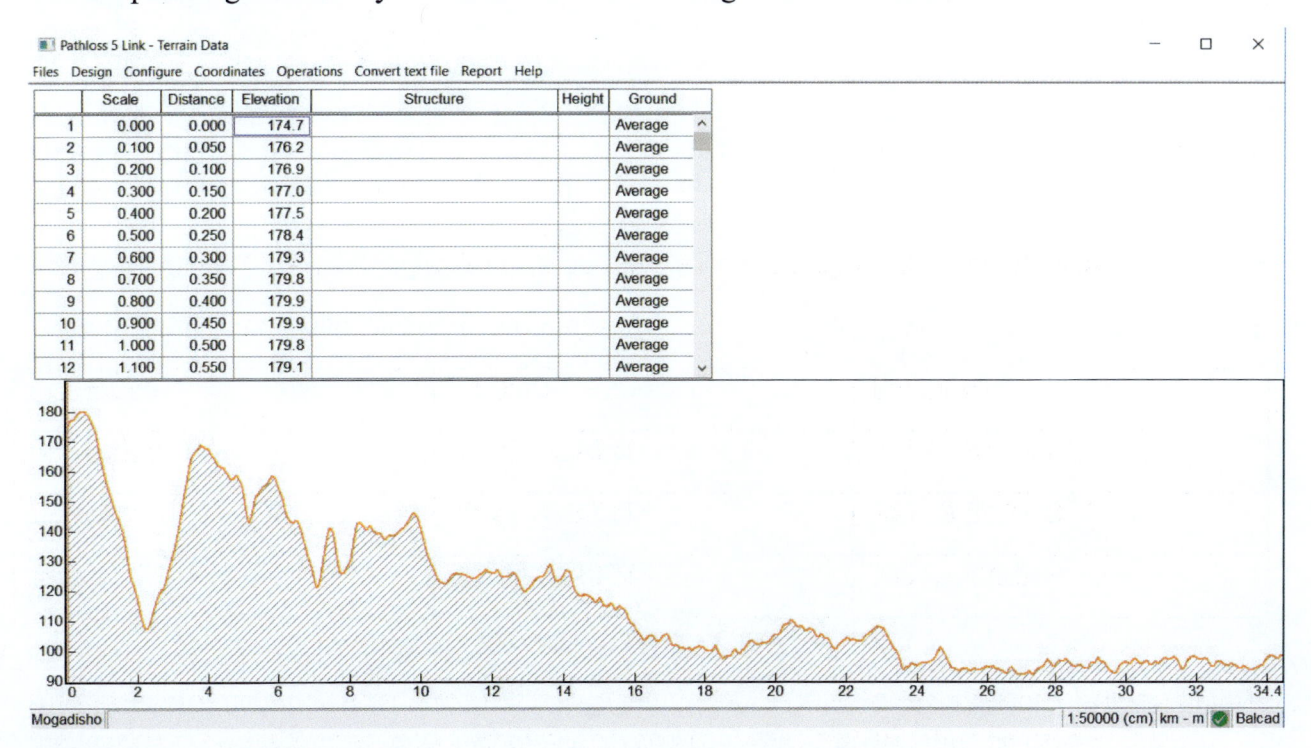

Figure 4.8 example of Pathloss 5.1 terrain profile between sites

As you can see, Mogadishu is at higher elevation than Balcad site, and the profile is gently falling as you go to Balcad. The distance between the two sites (hop length) is 34.4km

Route design

Next is to determine required tower height. To determine tower height, one needs to know that, as radio signals propagate across the ground, it is bent down (refracted) by atmosphere.

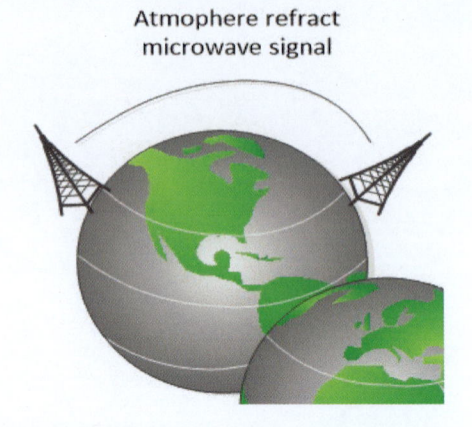

Figure 4.9 microwave terrestrial signals are refracted by atmosphere

The net effect is that microwave signals follow a curved path in the atmosphere. To get straight line path, the earth radius should be scaled by a factor called k-factor.

$$Earth\ propagation\ radius = k * (True\ earth\ radius)$$

The standard value of k used for propagation study is 4/3. Below this value the signal will bent downward to earth and hence diffraction loss. Above this value the signal is refracted upward and will result in atmospheric ducting.

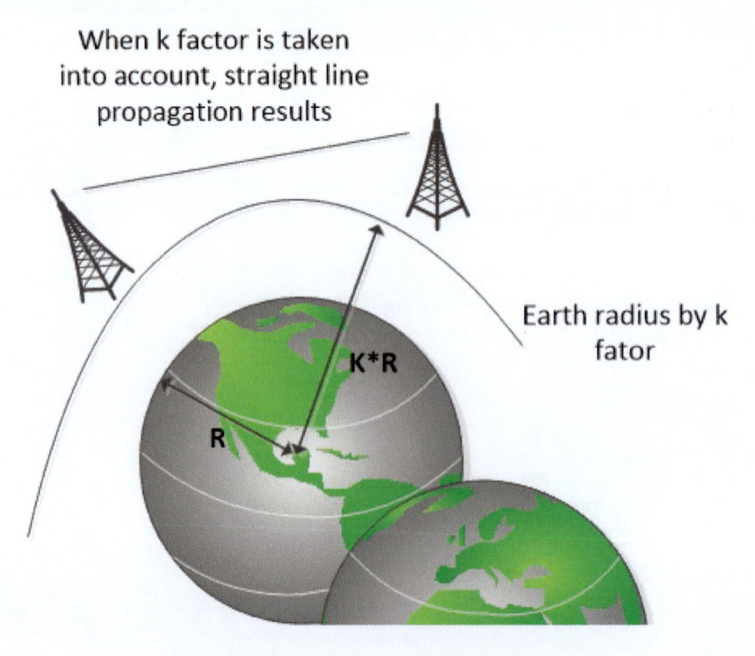

Figure 4.10 effect of k factor on propagation

When determining required tower heights, a particular value(s) of k-values is specified under line of sight clearance criteria.

If we choose k = 4/3 and our objective is 100% of first Fresnel zone to be un-obstructed, then the antenna heights will be

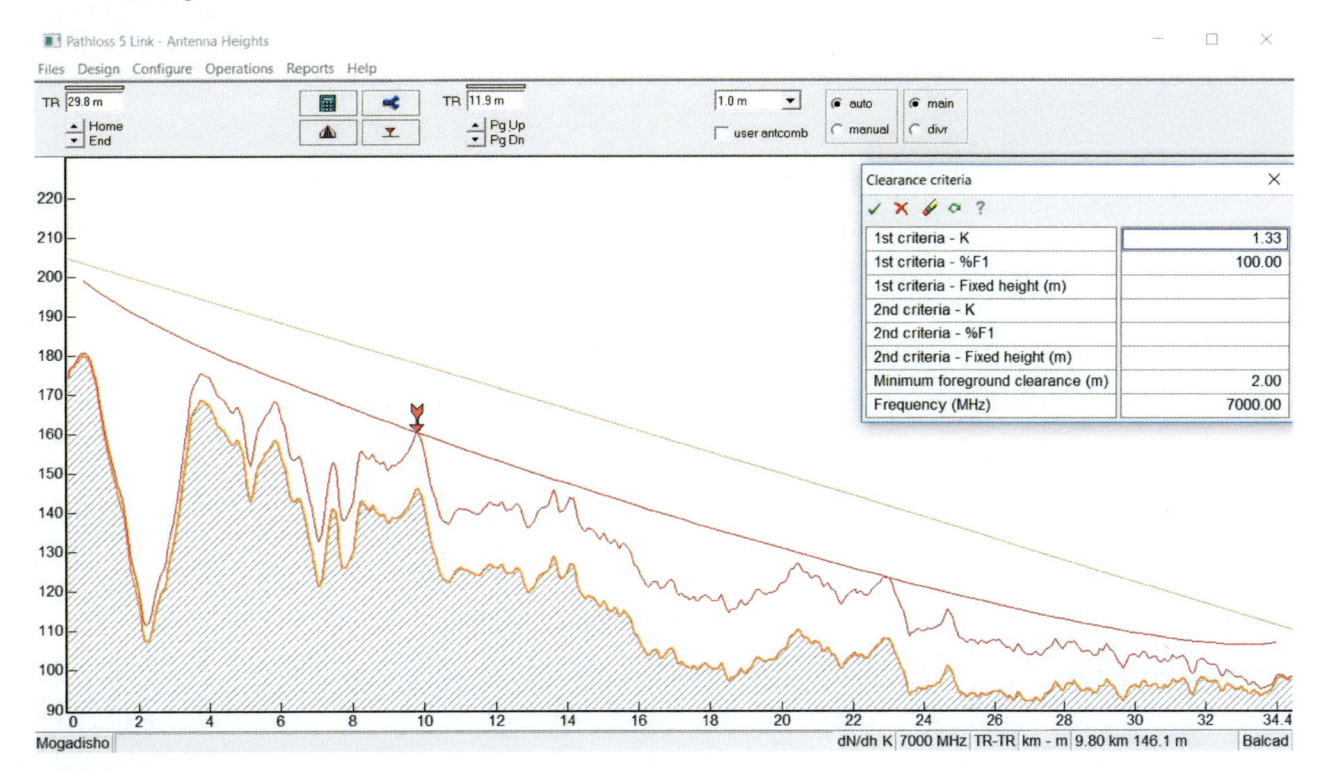

Figure 4.11 median k value effect

Mogadishu will have antenna height at 30m and Balcad at 12m.

However, k will vary with time. It will not stay at 4/3 for the whole duration. As k-value becomes smaller, diffraction loss will increase as shown below

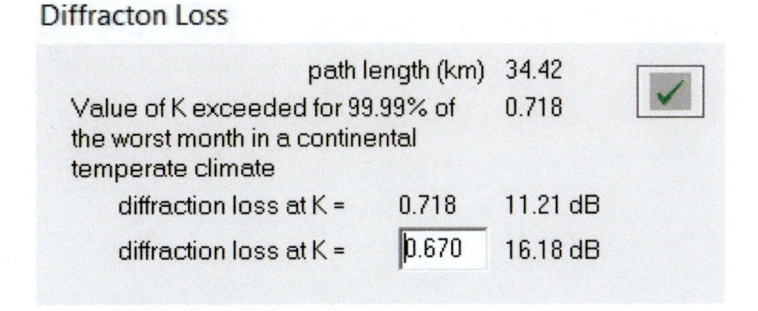

Figure 4.12 diffraction loss at median k value

As you can see, at k = 0.67 (2/3), diffraction loss is 16dB, and the link will perform poorly if link margin is not sufficient (say 40dB) to compensate for the diffraction loss

Now using the same antenna heights, let us modify the clearance criteria as follows

$100\% \; F1 \; @ \; k = \frac{4}{3}, \; and \; 30\% \; @ \; k = \frac{2}{3}$

This means at k = 4/3, 100% of first Fresnel zone (F1) will be visible between the sites, and at k = 2/3 only 30% of the first Fresnel zone is important.

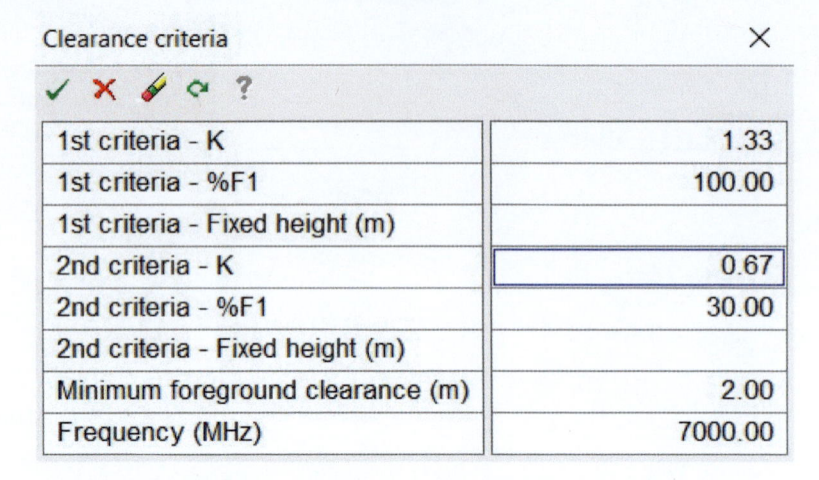

Clearance criteria	
1st criteria - K	1.33
1st criteria - %F1	100.00
1st criteria - Fixed height (m)	
2nd criteria - K	0.67
2nd criteria - %F1	30.00
2nd criteria - Fixed height (m)	
Minimum foreground clearance (m)	2.00
Frequency (MHz)	7000.00

Figure 4.13 adding minimum k to the path clearance criteria

The first clearance criteria (100% @ 4/3) is as shown above. The second clearance criteria (30% @ 2/3) will take into account sub-refraction affects due to diffraction loss at smaller k – value.

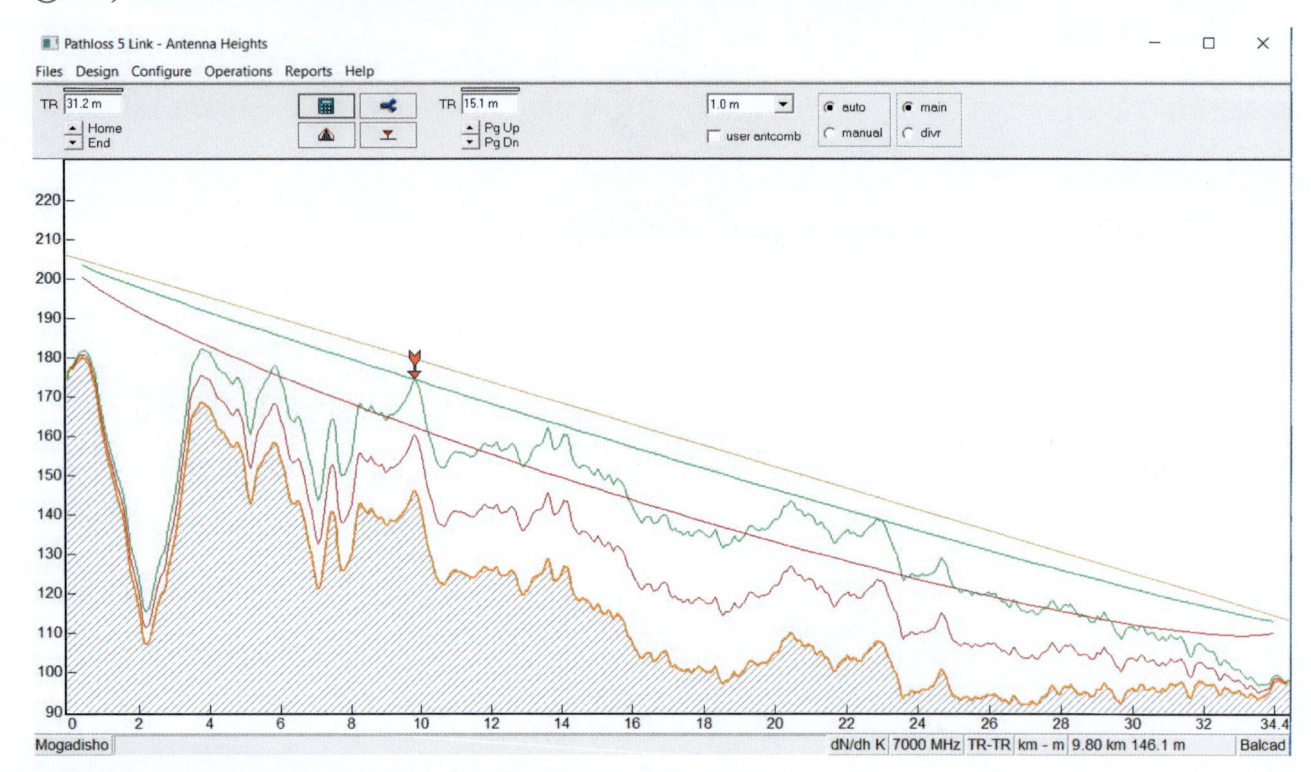

Figure 4.14 antenna height at median and minimum k values

When second clearance criteria is considered, antenna height now increases. The green earth and curve represent effective earth and Fresnel zone respectively at k = 2/3.

The red earth and red curve represent effective earth and first Fresnel zone respectively at k = 4/3.

To summarize

- Determine antenna height using standard k = 4/3 and 100% F1 clearance
- Calculate diffraction loss that will occur at smaller k-value say 2/3
- Increase antenna height so that diffraction loss at k = 2/3 can be safely ignored (much less than compared with link fade margin).

Please note that the introduction of second clearance criteria will reduce diffraction fading but also force increased antenna height (excess clearance) at median k = 4/3.

The second step of the LOS route design is to study the effect of this excess clearance. Excess clearance may cause signal reflection and multipath propagation. ITU-R P.530 gives detailed procedure on how to calculate antenna height when space diversity is used to overcome multipath.

We can use pathloss ray tracing under multipath-reflection analysis to implement this algorithm.

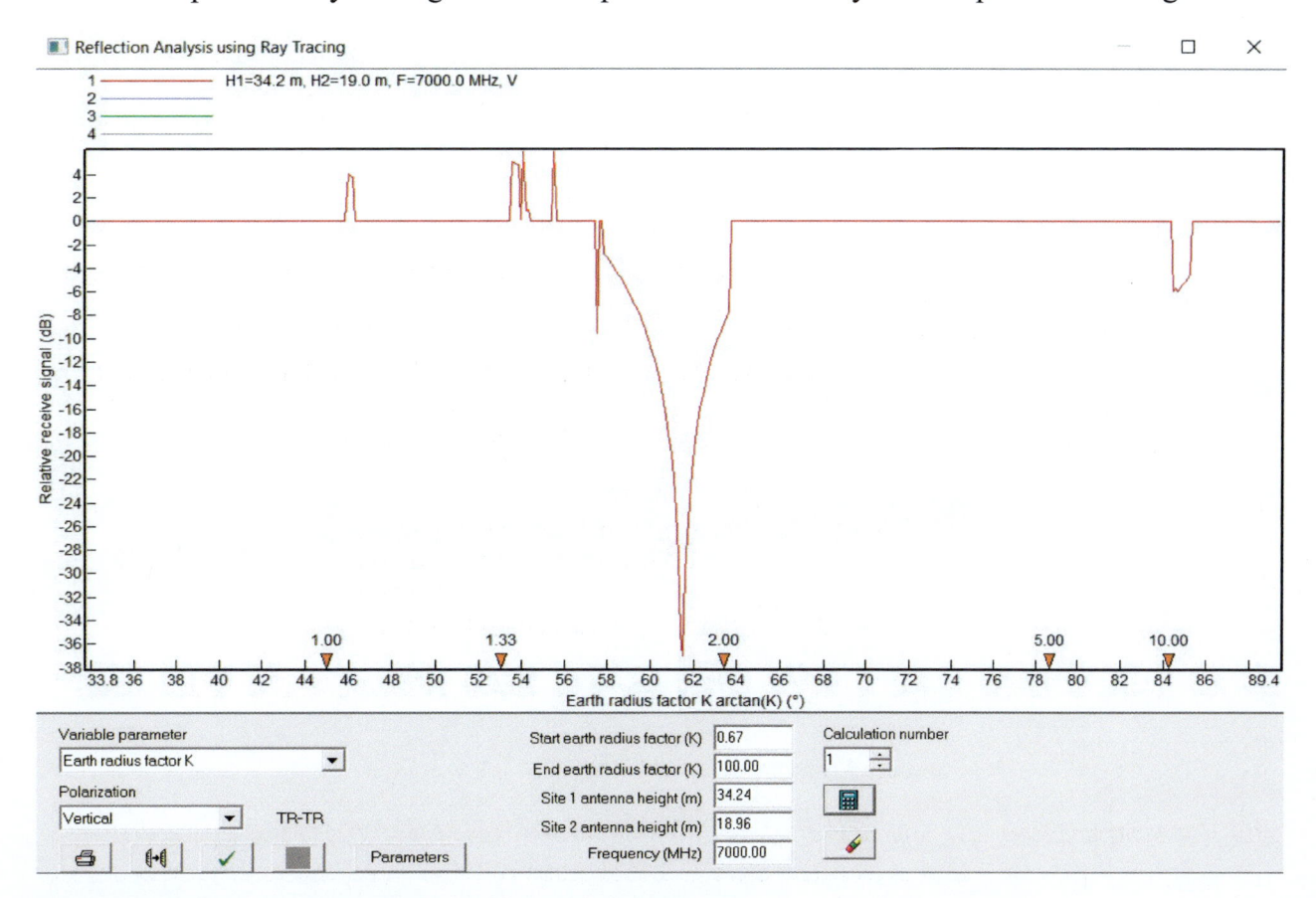

Figure 4.15 pathloss 5.1 multipath reflection at median k value

The figure shows that signal level is normal at median k = 4/3, but degrades as k-value approaches to 2.0 (super-refractive condition).

It would be advisable to implement space diversity in this link to overcome signal reflections. Space diversity uses two antennas on each side of the link. The top two antennas can transmit and receive. The bottom two antennas can only receive. The idea behind is that the two antennas on the site will not suffer signal nulls at the same time.

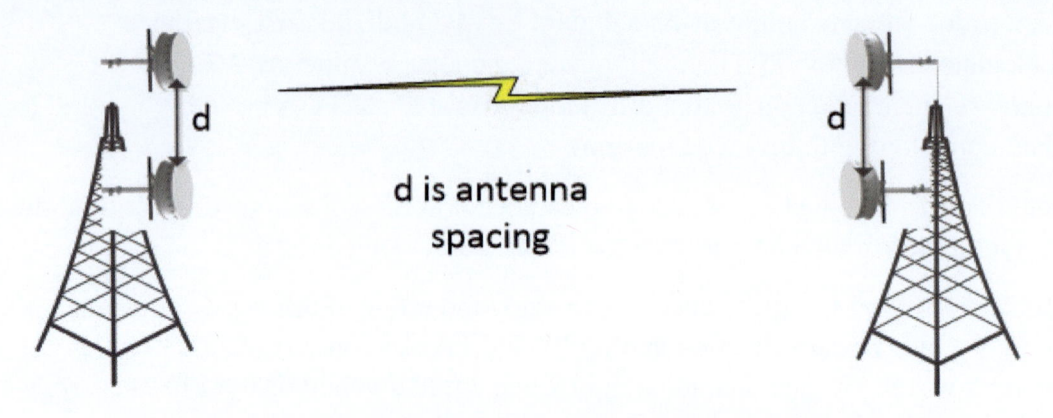

Figure 4.16 space diversity to combat multipath reflection

The space diversity antenna height should be 8 – 10m in spacing for towers that are not overcrowded

Chapter Five

Link design and installation

Link configuration and installation

From previous chapter, we seen concepts related to route design in which overall LOS connectivity is planned and radio towers built. Route design phase consumes significant time if resources are limited or ill-planned, it exposes to the operator huge overhead cost that will sometime go out of control.

If route design is finished, the next step is to plan individual links of the route one by one

Before we further go deeper into link design, let us present common microwave link configurations seen in practical network today

1 + 0 Basic microwave link

- Consists of single polarization antenna with one ODU on each end of the link
- It is called 1 + 0 as it does not have any protection

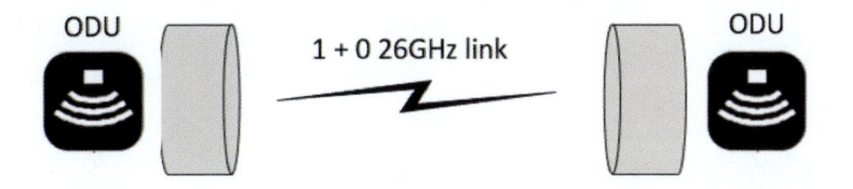

ODU directly Mounted antenna port

Figure 5.1 Basic unprotected microwave link

2 + 0 single polarization link

- Two ODUs are connected to the single polarization antenna via hybrid coupler. One ODU transmits at f1, while the other ODU at f2
- Two ODU directly mounted to antenna via coupler

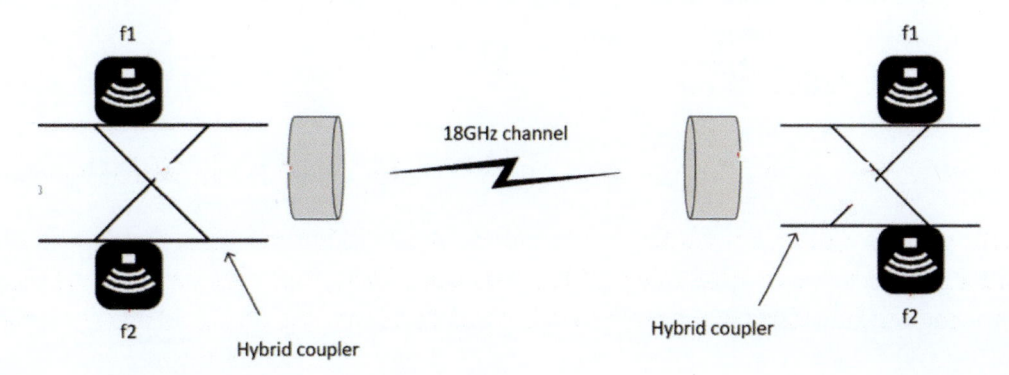

Figure 5.2 Single polarization protected with coupler

2 + 0 CCDP dual polarized link

- For dual polarized antenna, an OMT is required to directly couple two ODU with different polarization onto same antenna

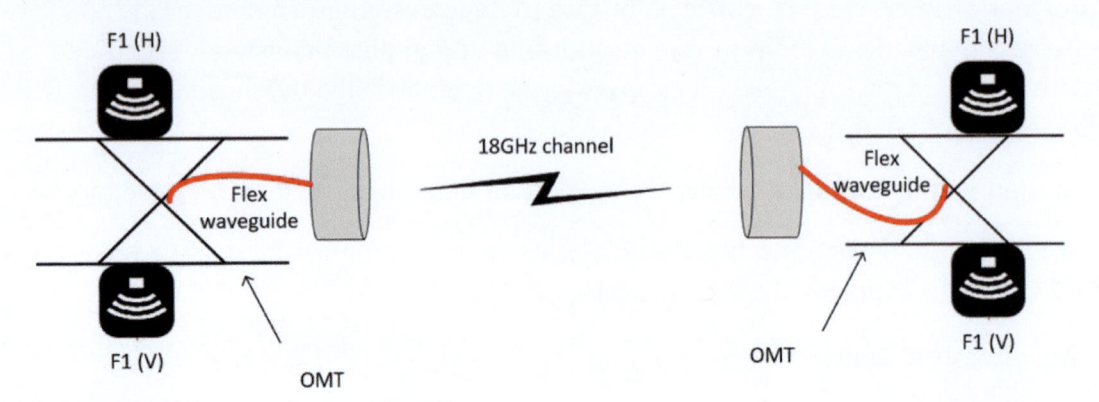

Figure 5.3 Dual polarized protected with OMT

1 + 1 Hot-standby single polarization link

- Two ODUs mounted onto the antenna transmit single frequency
- Only one ODU allowed to transmit while the other on standby
- This arrangement provides equipment protection

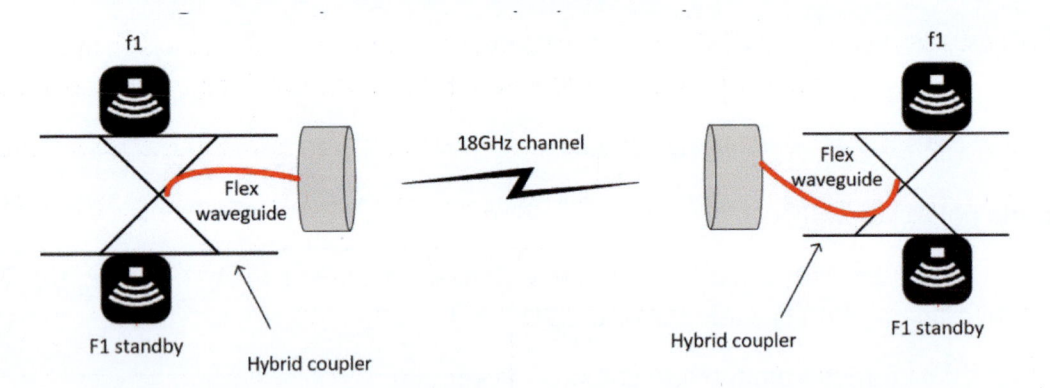

Figure 5.4 hot standby equipment protection

2 + 0 Space diversity design

On flat surface and water bodies, reflection become major cause of link failures.

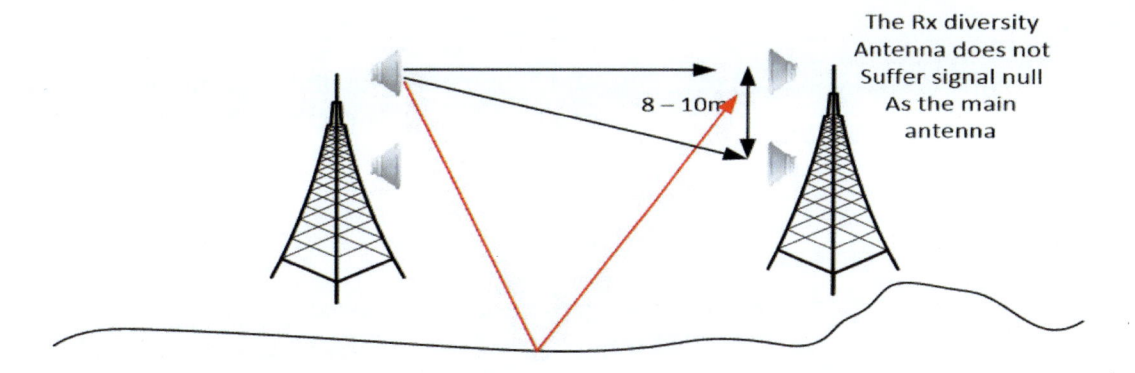

The Rx diversity Antenna does not Suffer signal null As the main antenna

8 – 10m

Figure 5.5 2+0 space diversity protection

Link design and installation

In space diversity top antenna transmits and receives while the lower antenna only receives. As shown in the figure above the red link is the reflected signal. On the right side the top antenna receive both main and reflected signal and cancel each other. With space diversity configured on the link, the lower antenna will save the link as it does only receives main signal without reflection.

In microwave link design you need to decide which antenna type and radio model is selected for equipment procurement and bill of quantity (BOQ) development

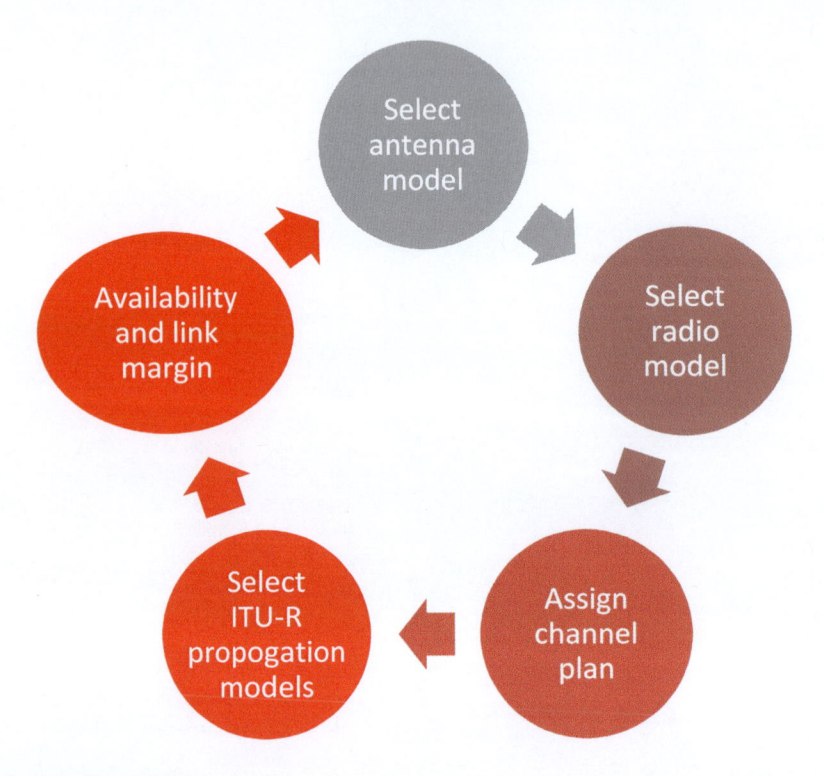

Figure 5.6 Link design flowchart

To gain real world experience, let us take a discussion meeting between management and technical team.

Discussion meeting between management
And technical team

Link design and installation

Today the management has sat down with the technical team to prepare design of new 25 links that will be added to the existing one as a part of capacity expansion. The manage asked the technical team the following questions

1. **Do we have links plan info** ?(coordinates, tower type and height, coverage zones, link length and current status)
2. **What are links capacity requirement and performance objectives**? (500mbps capacity and 99.9% availability is decided)
3. **Do we have available channel plan for this links for this links or we can re-use our existing spectrum**? (First it was discussed to apply for new spectrum license. However one engineer added that if we deploy the new links to the same location as the existing ones, it will increase capacity but still we don't have redundancy incase power failure. Then it was later decided to plan new route for this links as re-use the existing spectrum)
4. **What antenna size and model is suitable for these links**? (After pathloss link design, the link length was found to be long. Antenna of 2.4m was decided with higher gain)
5. **What are the features of the radio we need**? (after pathloss link design, HP radio with dual core and 112Mhz and 2048QAM is needed in addition with supporting 10G electrical and optical ports)
6. **Which vendor to go**? (technical team told the management to give them 1 week in order to negotiate with multiple vendors from both performance and cost perspective)

The good side about this dialog is that it happens every day in telecom operators. The bad side is that it is not easy as it is simply written for end products to perfectly reflect on the original plan

The next day the technical team sits down to do his duty and translate management requirement into practical realization of fully functioning microwave transmission network. In microwave transmission link design a lot of operators use Pathloss.

A typical transmission plan will open his laptop installed with Pathloss and performs the following steps

- Create new project by importing DEM SRTM 1arc into the program
- Site data entry either manually or import .csv site file
- Import antenna and radio file that will based on potential purchase
- Select ITU-R recommendations for propagation modeling
- Assigned radio channel with polarization (either vertical or horizontal) to the link
- Run the analysis and check performance output
- Make final decision if performance and reliability objectives has been achieved. Otherwise change the design and see if there is improvement.
- Print the design sheet and produce report
- Contact and negotiate with multiple vendors by receiving data sheet and quotation of their products

Next let us take a look at an example of typical link budget. The link budget is the ultimate decider of the antenna and radio specifications.

Link design and installation

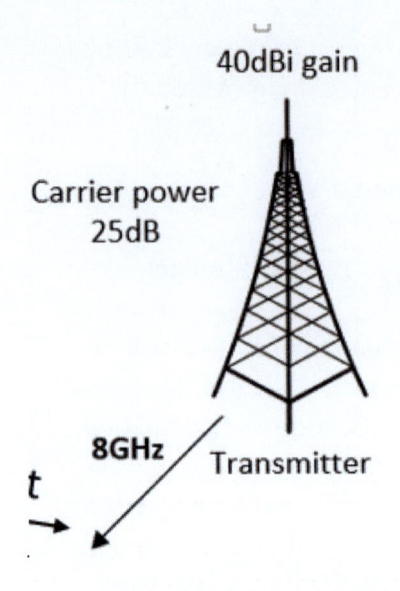

Tx power	25dB
Tx antenna gain	40dBi
System loss	3dB
Hop length	30km@8GHz
Free space loss	80dB
Rx level	-48dB
Rx sensitivity	-80dBm
Fade margin	32dB

Figure 5.8 Example link budget

As can be seen from the link budget, an antenna model is selected for this link with gain of 40dBi. The radio transmit power is 25dB transmitting at 8 GHz band. The main concept of the link budget is the receive signal level which -48dBm. For the link reliability to increase against fading and rain outage, the fade margin should be sufficient (>15-dB for good designed links)

With the sufficient link margin of 32dB for this link, the reliability will be optimum throughout the year.

If the link margin is found to be small (tends to be smaller for XPIC links), we can use antenna type of higher XPD. The effect of high performance antenna with high XPD is shown below

Link design and installation

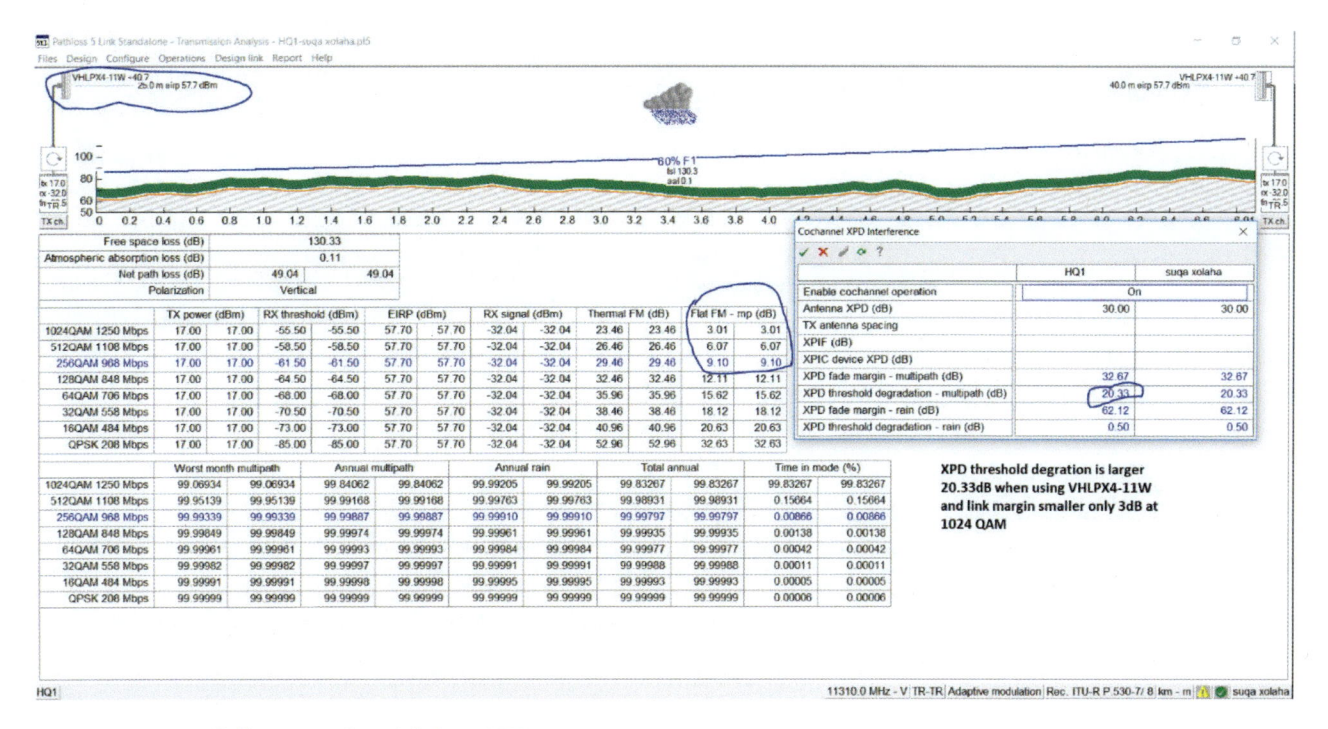

Figure 5.9 small flat margin with low XPD antenna

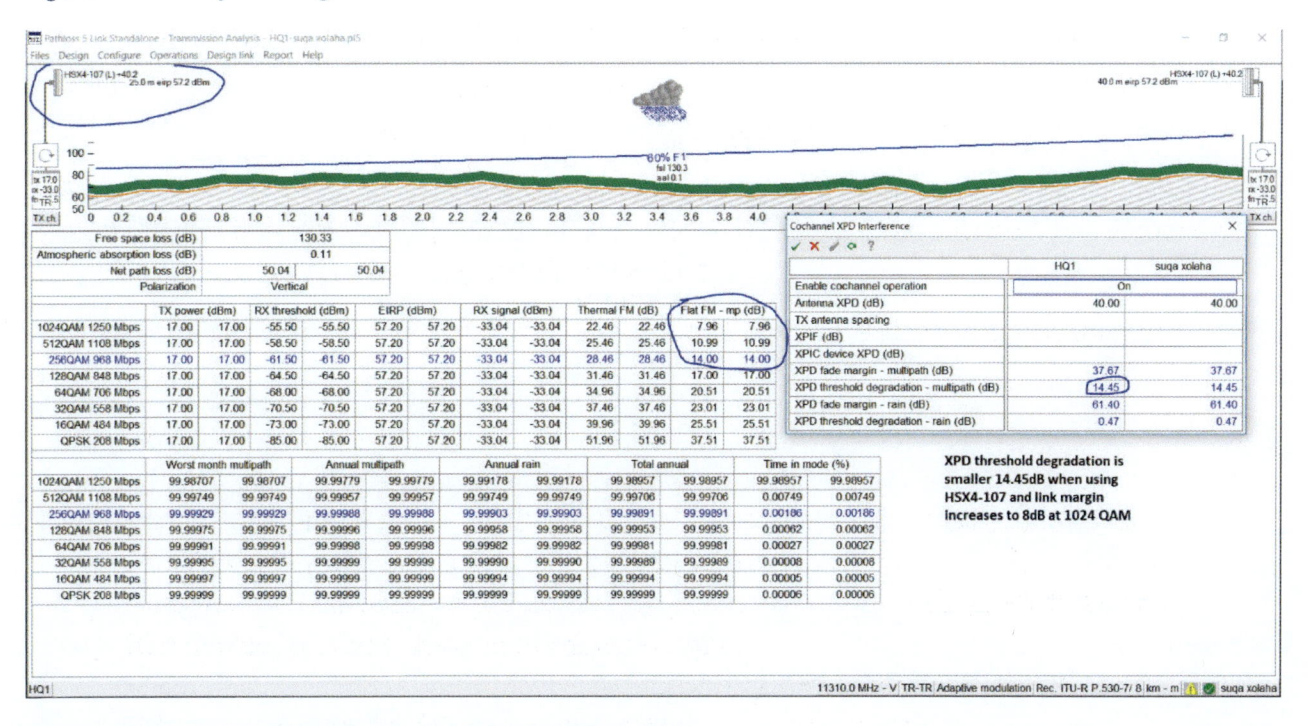

Figure 5.10 improved flat margin with high XPD antenna

As can be seen with the use of HSX antenna model which is higher XPD (40Db), the flat fade margin due to multipath improves.

What is the annual multipath and rain availability of the link?

- 99.99% availability (52.56min/year downtime)

53

- 99.9% availability (8.76hr / year downtime)
- 99% availability (3.65 days / year downtime)

How do we that with 99% availability the predicted outage per year is 3.65days due to rain and multipath fading?

Well in the year we have total of 365days. Now 99% of 365 is 361.35 with the remaining 3.65days being the outage expected throughout the year.

The above example touched multipath fading. However causes of radio fade are broad and include

- Fades produced by rain especially at higher frequencies (> 10Ghz)
- Diffraction fade coming from sub-refraction at lower k
- Radio fading due specular reflection in which main and reflected signals add to produce signal null
- There are also fades caused by surface and elevated ducts. An example of link that is free from surface duct (the wave trapped between the earth surface and the atmosphere) is shown in figure 5.11 below

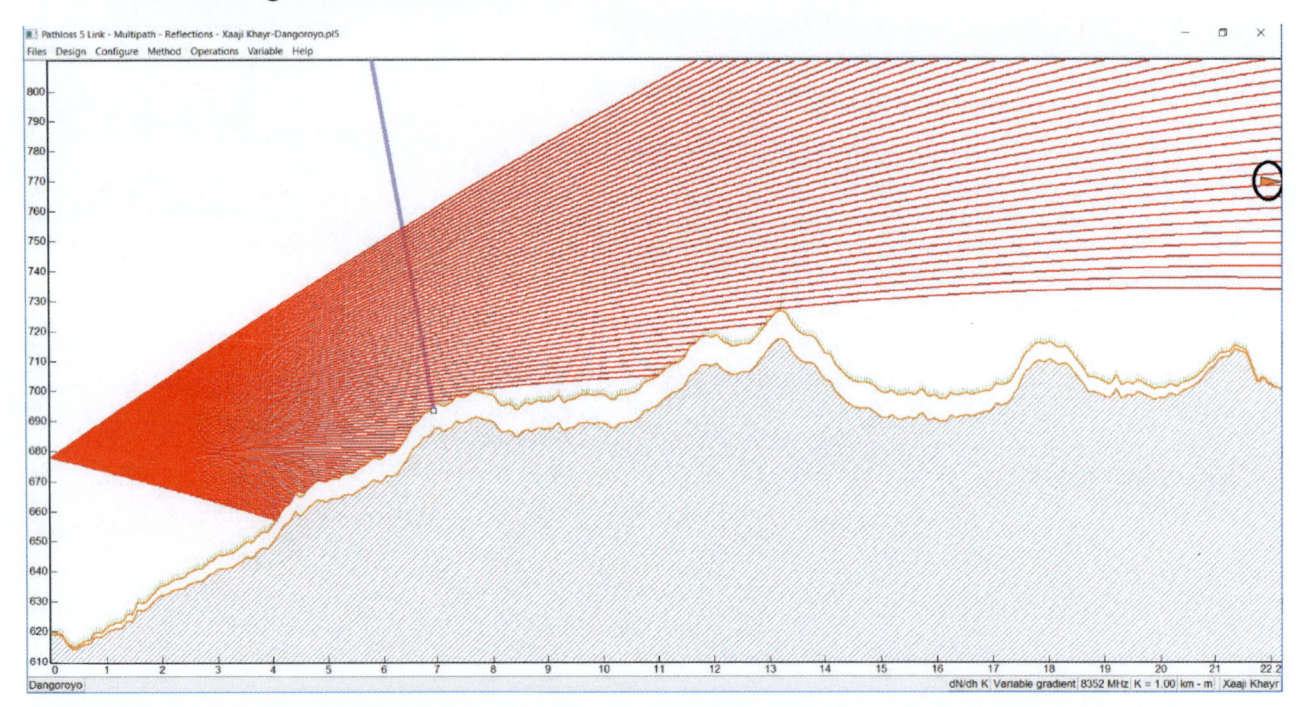

Figure 5.11 an example of link that is free from surface duct

In radio link design, each of these radio fades contribute to the overall performance degradation.

Now final meeting will be held to wrap up the design and select vendor. The operator manager now orders the equipment and after months shipped and arrives. Then what is next?

The microwave radio engineer will now have three more assignments in hand

- Link installation following best installation practices
- Link configuration and payload placement
- Link monitoring and maintenance
- Link upgrades

Installation phase

Installation will start right after link design is finalized and equipment arrives. Supply chain department will play role in coordinating equipment dispatch to sites.

Once engineers and their supervisor reach the site, they will first do visual inspection and count everything. Instructions are explained from equipment data sheets

In modern times, operators are pushing for all-outdoor installation to minimize indoor air conditioners cost. In such case all radio equipment will be installed on the tower and single Ethernet cable carrying both power and data will descend to the equipment room. The disadvantage is all your equipment is not sheltered, and hence stray bullets can damage in sensitive areas.

The scenario of single Ethernet cable for both data and power works for access sites in metropolitan areas where site access by troubleshooting team is very fast as shown in figure 5.12 below

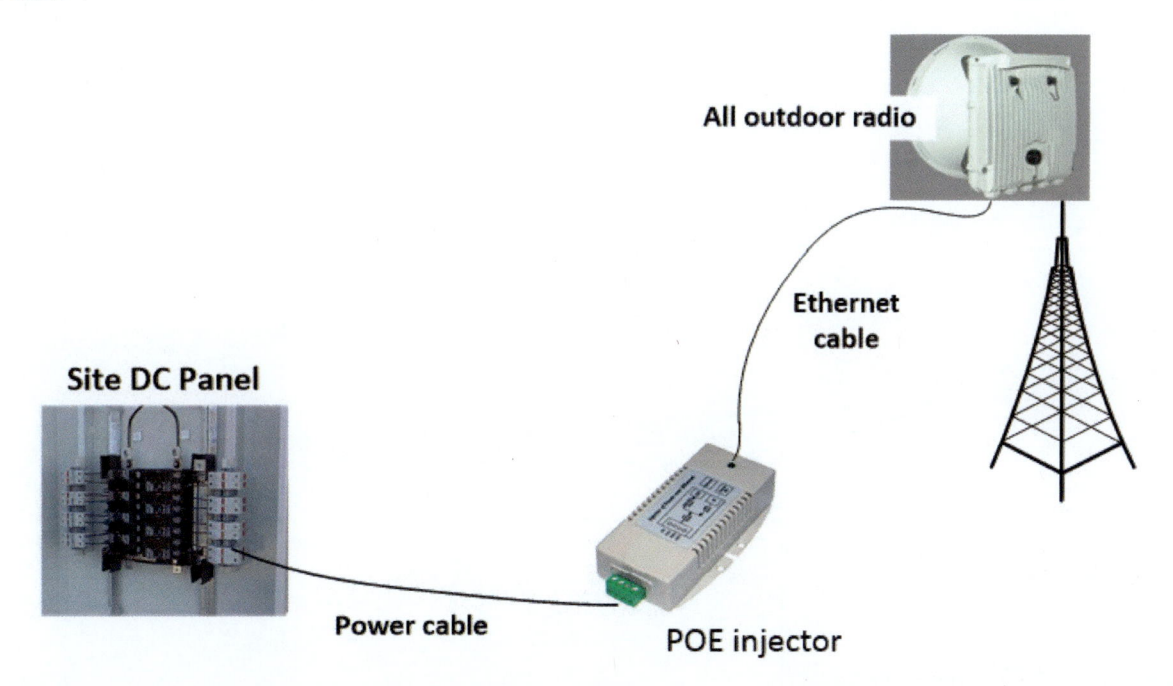

Figure 5.12 all outdoor single cable for power and data

For remote backbone sites where engineers are far away to troubleshoot, it is safer to use separate power and data cables. POE injector is not needed in this case to invest. This is shown in figure 5.13 below

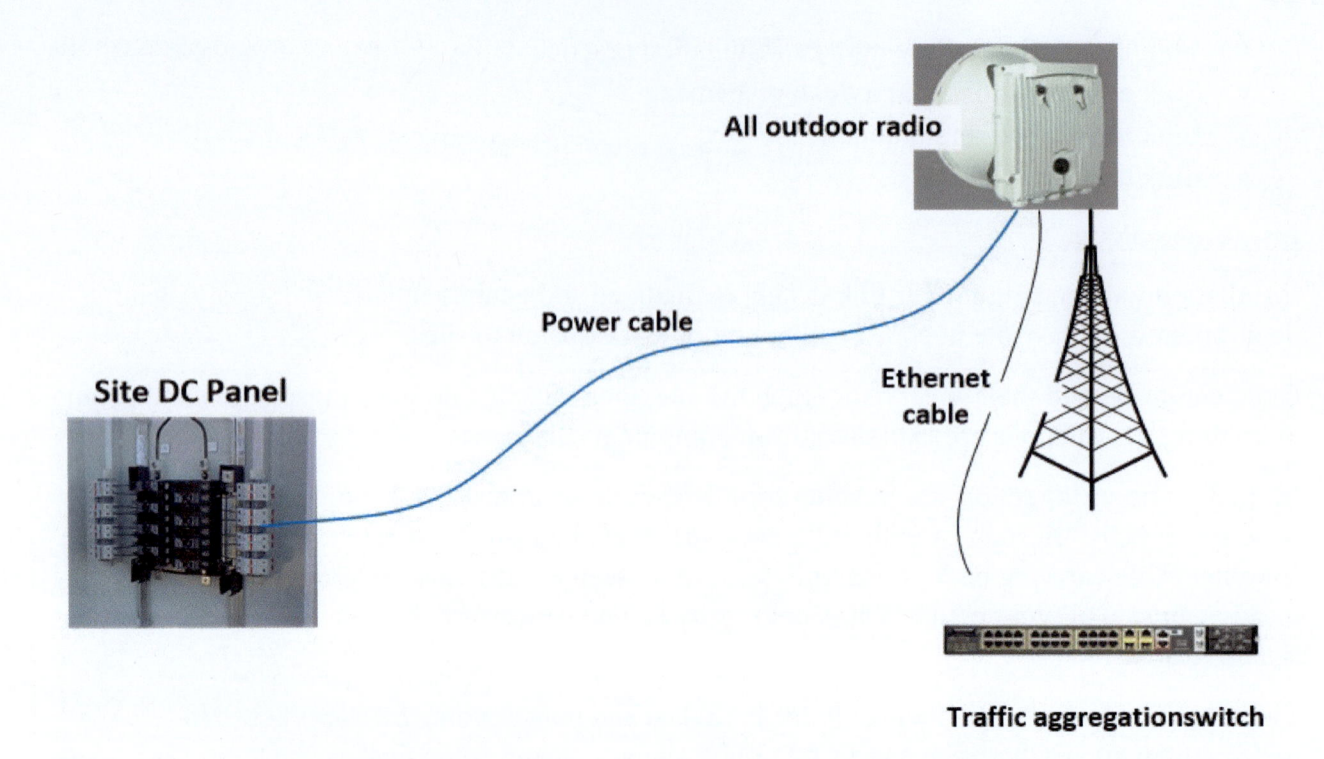

Figure 5.13 all outdoor separate cables for power and data

During installation various cautions must be taken as explained in vendor equipment installation manual. Proper training of technician is mandatory as well as team work capacity building. A typical worksheet to follow as a guideline in installation and final auditing is shown in the table 5.1 below

Table 5.1 link installation validation checklist

Site name		
Installation time (days)		
Antenna size (m)		
Antenna height (m)		
Number of antenna on the tower		
Is antenna mounting secure?		
Is the mount properly grounded?		
Is the antenna properly grounded?	Yes ☐	No ☐
Are all antenna and ODU connector point weather-proofed?		
Is IDU properly mounted on rack?	Yes ☐	No ☐
Sufficient space for ventilation from other devices on the rack?	Yes ☐	No ☐
Is IDU properly grounded on the rack?	Yes ☐	No ☐
Are all cables properly labeled and secure?	Yes ☐	No ☐
Is the fan working?	Yes ☐	No ☐
Is the rack properly grounded?	Yes ☐	No ☐
Redundant power to IDU	Yes ☐	No ☐
Measured input voltage to IDU		
Is the DC connector secure via the fasten screws?	Yes ☐	No ☐
Number of IF cables on the tower		

Average cable length (m)		
Number of surge arrestors installed		
Is cable connectors weather-proofed?	Yes ☐	No ☐
Number of ODUs on the tower?		
Is polarization same on both ends of the link?	Yes ☐	No ☐
What is the polarization?		
Is XPD value between recommended 25 – 35	Yes ☐	No ☐
Commissioned RSL values (dBm)		

One important thing to note is as transmission team head to the sites where the link is to be installed, they need these two information to be with them from planning team

- Equipment data sheets (normally included inside the equipment box)
- Link design file

An example of link design file planned in Pathloss software is shown in tables below

Table 5.2 link design output in Pathloss 5.1

	A	B
Latitude	**Latitude**	**Latitude**
Longitude	**Longitude**	**Longitude**
Easting (m)	537965.5	539123.7
Northing (m)	225852.9	226938.2
UTM zone	38N	38N
True azimuth (°)	46.87	226.87
Vertical angle (°)	0.73	-0.74
Elevation (m)	37.02	52.19
Tower height (m)	**30.00**	**30.00**
Antenna model	**VHLP1-23 (TR)**	**VHLP1-23 (TR)**
Antenna file name	7172	7172
Antenna gain (dBi)	35.30	35.30
Antenna height (m)	**10.53**	**15.77**
Frequency (MHz)	22652.00	
Polarization	**Vertical**	

	A	B
Path length (km)	1.59	
Free space loss (dB)	123.59	
Atmospheric absorption loss (dB)	0.31	
Net path loss (dB)	53.29	53.29
Radio model	**Radio X**	**Radio X**
Radio file name		
Emission designator		
TX channel assignments	**15h 23268.00V**	**15l 22036.00V**
Geoclimatic factor	1.115E-005	
Path inclination (mr)	12.85	
Average annual temperature (°C)	10.00	
Fade occurrence factor (Po)	2.389E-007	
Polarization	Vertical	

Table 5.3 ACM link RSL and margin performance

	TX power (dBm)		RX threshold level (dBm)		EIRP (dBm)		Receive signal (dBm)		Thermal fade margin (dB)		Flat fade margin - multipath (dB)	
1024QAM 460 Mbps	20.00	20.00	-56.50	-56.50	55.30	55.30	-33.29	-33.29	23.21	23.21	23.21	23.21
512QAM 417 Mbps	20.00	20.00	-59.50	-59.50	55.30	55.30	-33.29	-33.29	26.21	26.21	26.21	26.21
256QAM 371 Mbps	20.00	20.00	-62.50	-62.50	55.30	55.30	-33.29	-33.29	29.21	29.21	29.21	29.21
128QAM 321 Mbps	20.00	20.00	-65.50	-65.50	55.30	55.30	-33.29	-33.29	32.21	32.21	32.21	32.21
64QAM 250 Mbps	20.00	20.00	-69.50	-69.50	55.30	55.30	-33.29	-33.29	36.21	36.21	36.21	36.21
32QAM 191 Mbps	20.00	20.00	-72.50	-72.50	55.30	55.30	-33.29	-33.29	39.21	39.21	39.21	39.21
16QAM 150 Mbps	20.00	20.00	-76.00	-76.00	55.30	55.30	-33.29	-33.29	42.71	42.71	42.71	42.71
QPSK 75 Mbps	20.00	20.00	-87.00	-87.00	55.30	55.30	-33.29	-33.29	53.71	53.71	53.71	53.71

Table 5.4 ACM link availability performance

	Worst month multipath		Annual multipath		Annual rain		Total annual		Time in mode (%)	
1024QAM 460 Mbps	99.9999	99.9999	99.9999	99.9999	99.9999	99.9999	99.9999	99.9999	99.9999	99.9999
512QAM 417 Mbps	99.9999	99.9999	99.9999	99.9999	99.9999	99.9999	99.9999	99.9999	0.0000	0.0000
256QAM 371 Mbps	99.9999	99.9999	99.9999	99.9999	99.9999	99.9999	99.9999	99.9999	0.0000	0.0000

	Worst month multipath		Annual multipath		Annual rain		Total annual		Time in mode (%)	
128QAM 321 Mbps	99.9999	99.9999	99.9999	99.9999	99.9999	99.9999	99.9999	99.9999	0.0000	0.0000
64QAM 250 Mbps	99.9999	99.9999	99.9999	99.9999	99.9999	99.9999	99.9999	99.9999	0.0000	0.0000
32QAM 191 Mbps	99.9999	99.9999	99.9999	99.9999	99.9999	99.9999	99.9999	99.9999	0.0000	0.0000
16QAM 150 Mbps	99.9999	99.9999	99.9999	99.9999	99.9999	99.9999	99.9999	99.9999	0.0000	0.0000
QPSK 75 Mbps	99.9999	99.9999	99.9999	99.9999	99.9999	99.9999	99.9999	99.9999	0.0000	0.0000

Multipath fading method - Rec. ITU-R P.530-7/ 8
Rain fading method - Rec. ITU-R P.530-8/13 (R837-5)

To explain a little bit at the design file above generated from Pathloss software

- Sites A and B information provided such as (tower height, coordinates, required antenna height)
- The design says that antenna size suitable for this link is 0.3m using frequency band 23Ghz
- Required polarization to be installed in vertical
- TX channel duplex pairs is also provided to be configured in the radio (provided the area is already scanned for possible interference using interference analyzer)
- It also indicated the link will use ACM (adaptive coding and modulation) configuration in which link capacity will vary according to adaptive selection of modulation scheme based on radio propagation condition
- Under good propagation condition (in which best installation practice plays greater role), the operator modulation will be 1024-QAM that offers around 500Mbps when antenna pointing is fixed for receive level of -56dBm for each site and link margin of 23dBm is achieved. Things that may delay to achieve these results are installation or configuration error or site power failure or other natural circumstances such as rain.
- Under this design the link is expected to maintain those levels at 99.999% of the year.

When configuring channels or frequencies on the radio configuration page (accessed via radio default IP address and user credentials), pay in mind that microwave radio link are full duplex. That mean both site will transmit and receive at the same time. If the transmission frequency of site A is f1, this will be configured as receive frequency of site B and vice versa. This is shown in figure 5.14 below

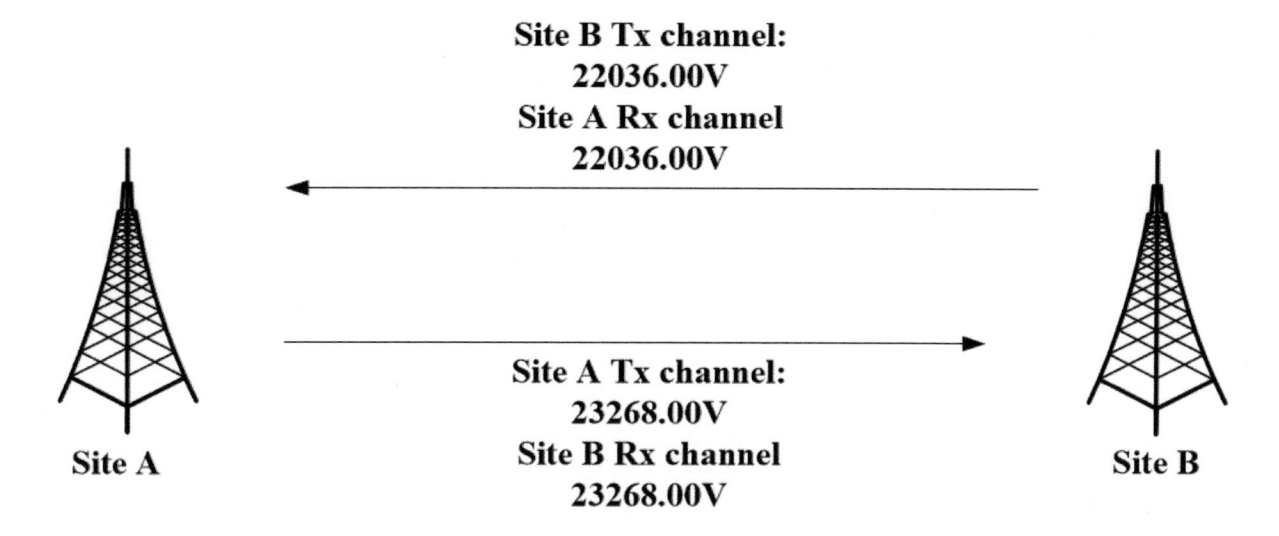

Site B Tx channel:
22036.00V
Site A Rx channel
22036.00V

Site A Tx channel:
23268.00V
Site B Rx channel
23268.00V

Site A   Site B

Site A Tx is greater than Site B Tx
Hence Site A is high site and Site B is low site

Avoid high/low violation

Figure 5.14 channel assignment for high and low stations

To sum up this section, the following table illustrates sections available in a typical radio modem configuration page.

Table 5.5 typical radio management portal

Product overview	
Device management	
Radio configuration	
IP configuration	
VLAN configuration	Configuration area for the engineer to click, select, apply, save and backup
Routing configuration	
Synchronization	
Spanning tree	
Ring protection	
CLI terminal	
ERPS	

Link design and installation

Performance	
Advanced settings	

In situations where higher capacity are required with no space diversity (SD) 4 + 0 configuration can be used with single OBU (outdoor branching unit) 4-port coupler. Using dual polarized antenna only 2 frequencies will be required.

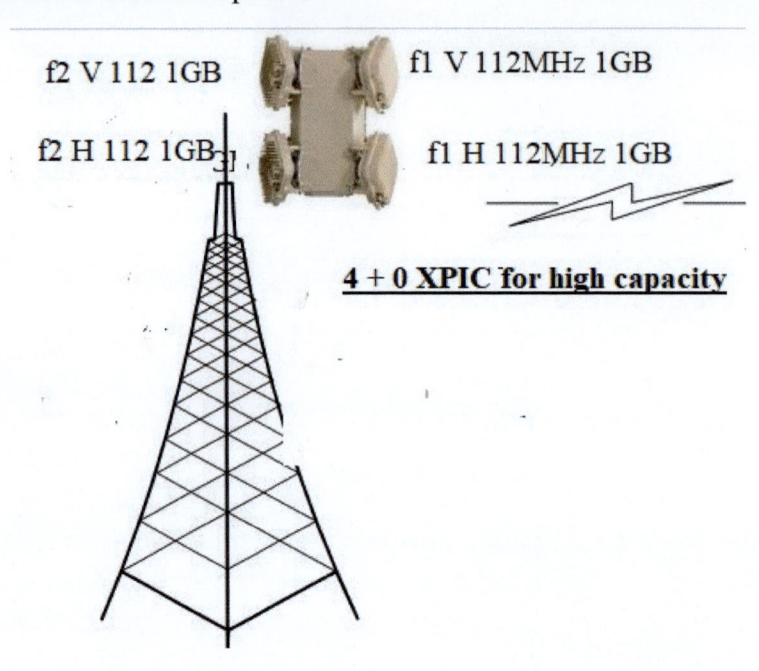

Figure 5.15 4 + 0 XPIC high capacity aggregation

In another deployment scenario 8 ODUs can be combined by dual polarized antenna in which each antenna houses one OBU. In such scenario 8GB microwave link is aggregated from multiple 1G independent radios for high capacity applications including 5G services, 8 radios each of capable of 1GB is aggregated by two OBUs. Using dual polarized antenna only 4 frequencies will be required in both polarization. Space diversity could also be used as well in this scenario. In such case you will have both path protection and link protection.

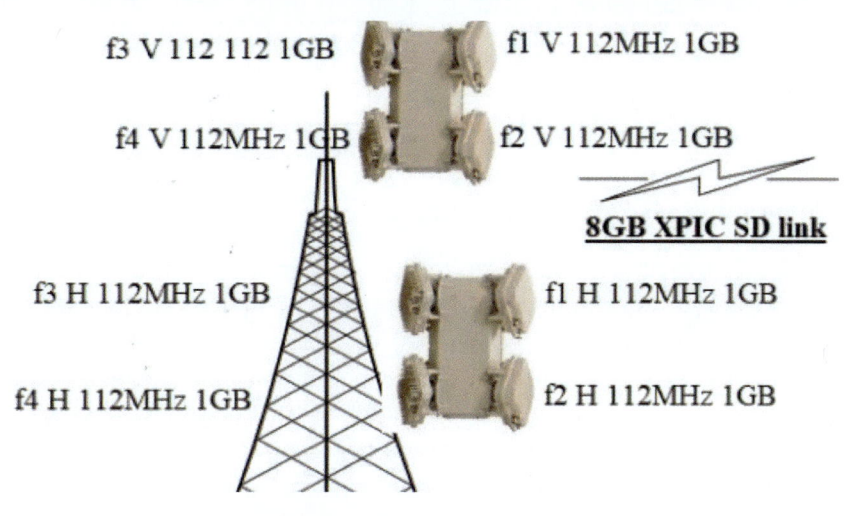

Figure 5.16 8+0 XPIC SD high capacity aggregation

Link traffic configuration and protection

So far we have discussed what is required from planning team to design microwave links and what are required during installation phase. Now the link is ready which means planning, installation and radio link is setup. The LEDs and monitoring tools all show everything is ok. The radio performance matches with planning data. And all physical connections are labeled, secured and tightened.

Well the operator didn't invest all that money the microwave link or links to be idle. Customer traffic should be terminated on the transmission network interfaces and VLANs configured.

We will take a look this part as follows

- Traffic configuration as VLANs
- Customer VLAN approach vs provider VLAN approach
- Redundancy and protection at link, network and payload levels
- Overall link capacity to fulfill required customer traffic such as cellular backhauling and internet services to businesses
- Interconnection to other network components such as mobile core network and data center switches
- Aggregation, add/drop and only pass-through sites need different requirements for traffic configuration
- Practice of avoiding single point failure as the telecom services have very high availability

Link design and installation

Before you start deploying microwave links the planning team should answer the following

- What is the capacity required between the transmission network and the mobile core network?
- What is the typical aggregation switch required and does it support higher capacity interface?
- Do we need to use Ethernet or multimode fiber to implement the interconnections?
- How are we going to manage the microwave links and what is IP and VLAN reserved for the network management server?
- What is purposed plan for network equipment protection and configuration backup?
- If an additional optical fiber link is available or leased from another operator, can we install and configure link protection using the ERPS feature available in the radio?

Different operator take different approach depending on simplicity and cost. At the end of the day the transmission network may interconnect two RNC located in different cities or two data centers located few kilometers apart.

The following figure shows one such approach taken to interface the transmission network with other parts of the overall operator network. A single high capacity optical fiber interconnect the transmission aggregation switch and the core network router and both traffic and management payload pass through single medium. Loops may arise if same VLAN is used for both management and traffic or incorrect IP plans are implemented

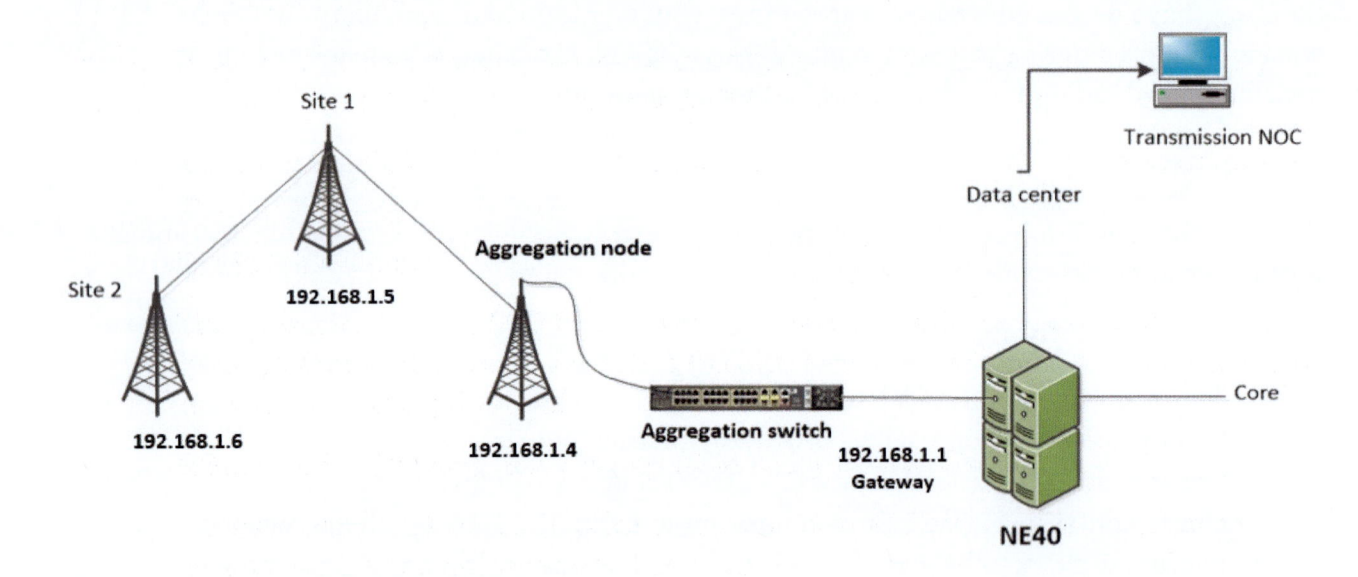

Figure 5.17 single connection between transmission and core for management and traffic

A second approach safer in which customer traffic such as mobile traffic goes to core network on a particular interface of the switch and management traffic goes to a separate firewall located in the data center room. In this way traffic and management are separated.

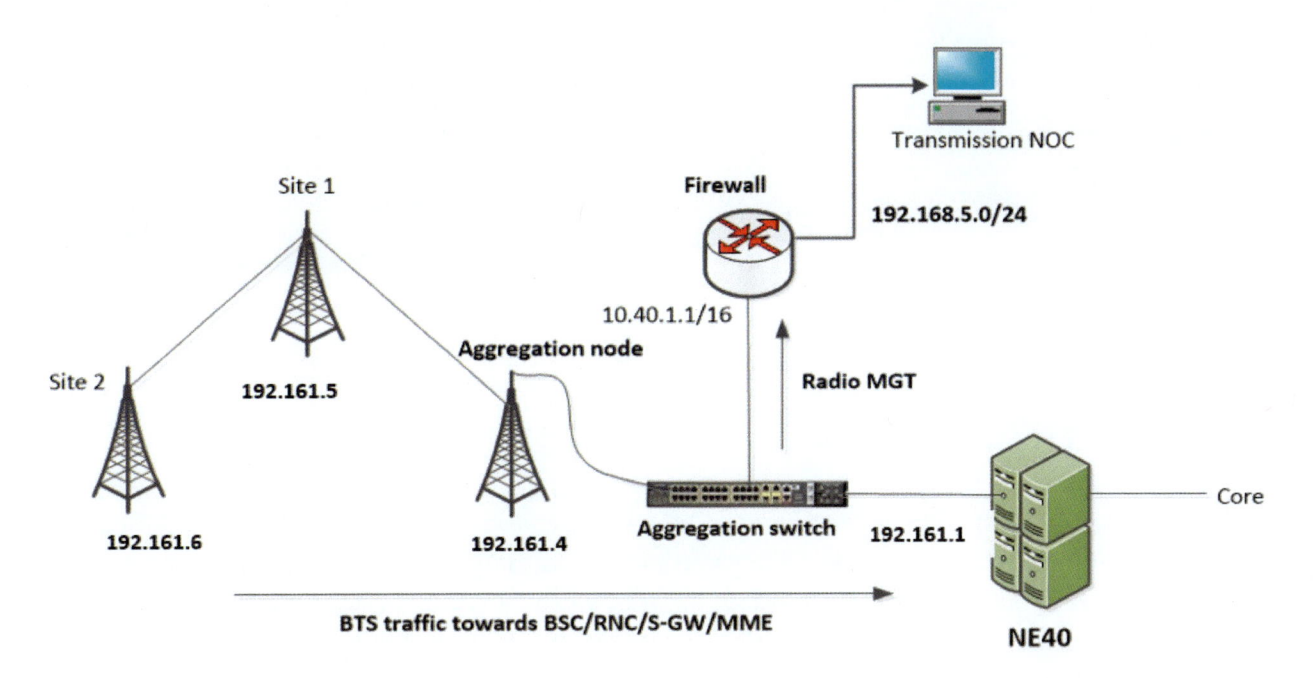

Figure 5.18 separate connection for traffic to core and management to firewall

We are assuming Huawei core network in this case. The core edge route is called NE40 and one of its 10G ports should be connected to the transmission aggregation switch 10G port by optical fiber. To avoid single point failure, a second standby connection is established. Normally the connection is IP in which the NE40 routed port acts as gateway to BTS sites. Thus a wireless engineer using U2000 core network manager can remotely access the BTS VLAN interface IP. The aggregation switch should have another 10G port in the uplink radio path to aggregate all traffic coming all remote sites.

The transmission access switch is then interfaced to uplink radio path on the tower. CAT 5E/6 is run from the equipment shelter where the aggregation switch is located all the way to the radio 10G port.

An example of transmission aggregation used in typical production operator network is shown below

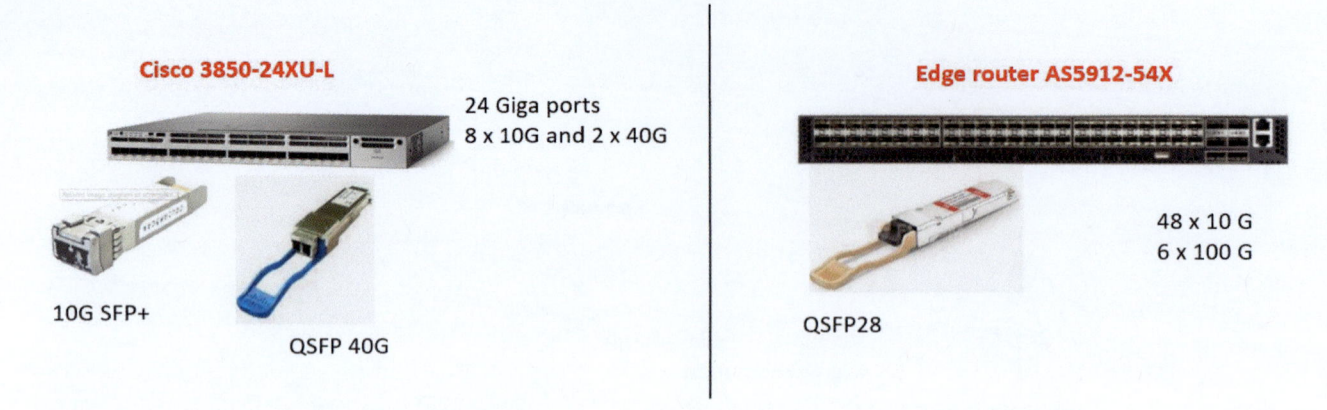

Figure 5.19 typical aggregation switches

If a customer wants Ethernet traffic to be transported from HQ to a client office, VLAN should be configured for that service in the aggregation switch, indoor modem, and radio on both ends of the link. To minimize human error, the radio modem is configured as provider bridge mode. What this means is that the radio port facing to aggregation switch is configured as QinQ instead of transparent bridge. This allows no VLAN to be configured on the radios except the provider VLAN. The design layout is shown below

Link design and installation

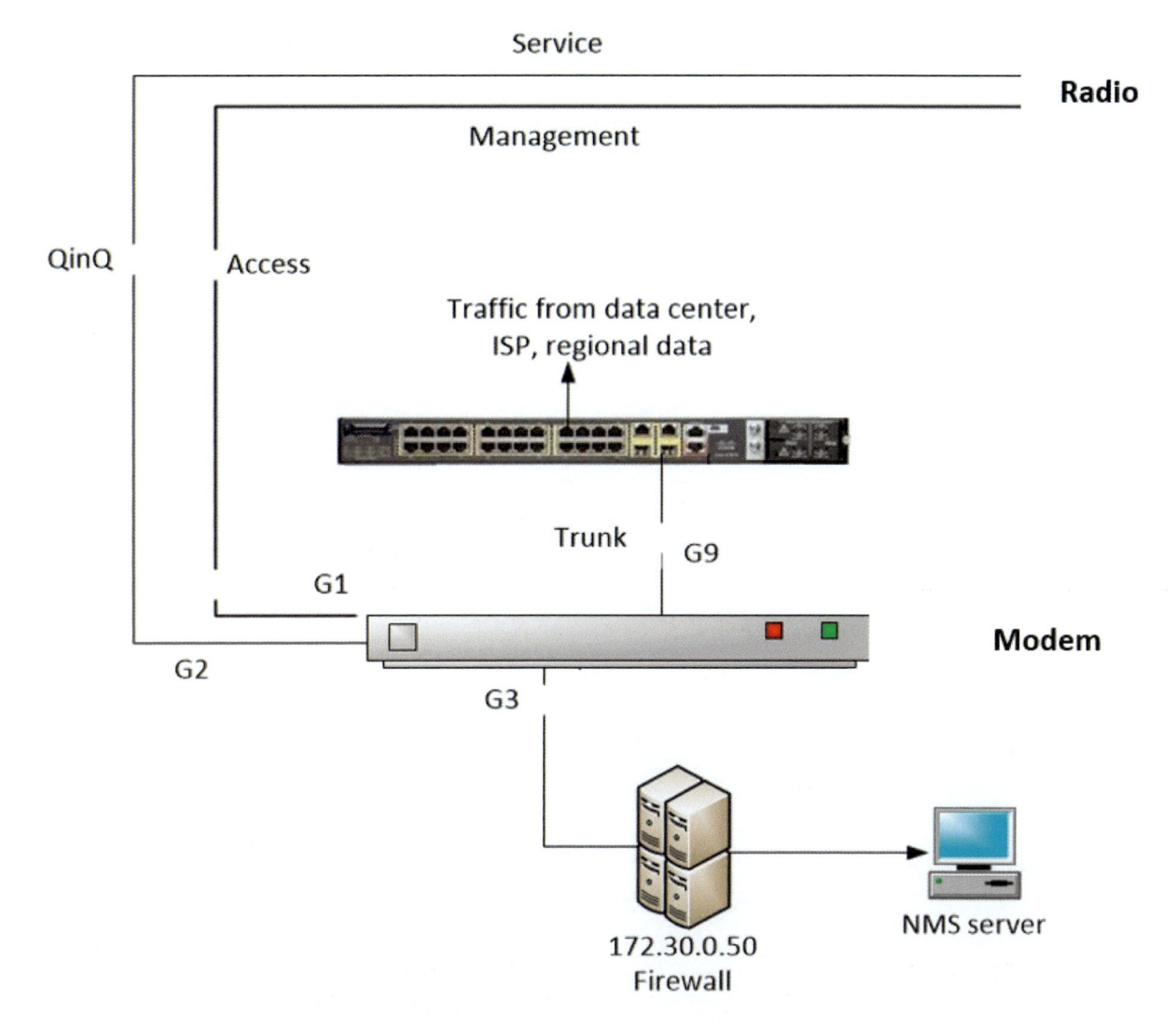

Figure 5.20 typical aggregation center and using QinQ double tagging using all outdoor radio

As an example let us illustrate a typical configuration of the above network as follows

For aggregation modem we have as

Table 5.6 VLAN configuration example

VLAN ID	NAME	MEMBER PORT	DESCRIPTION	MODE
100	10Mbps internet to certain hotel	G9	Port facing to aggregation switch	Trunk
100	10Mbps	GX	Port facing to ISP rack	access

For aggregation switch

Table 5.7 VLAN configuration example

VLAN ID	NAME	MEMBER PORT	DESCRIPTION	MODE
100	10Mbps internet to certain hotel	G9	Port facing to aggregation switch	Trunk
100	10Mbps internet to certain hotel	G2	Port facing to radio	Trunk

For radio on tower

Table 5.8 VLAN configuration example

VLAN ID	NAME	MEMBER PORT	DESCRIPTION	MODE
1000	10Mbps internet to certain hotel	P1	Port facing to aggregation modem	QinQ

What this example says is that when traffic leaves from ISP switch as VLAN 100 and enters the transmission network, it will be wrapped in another VLAN 1000 using the Ethernet protocol QinQ licensed in the radio. In this way customer VLAN 100 is preserved as it moves through the radio chain. At the other side of the link same configuration is applied and VLAN 100 is set free at radio egress port towards the hotel local switch for internet service.

This examples has only shown single link configuration for traffic. What about for daisy chained links where the client location is a few hops away. In such case the pass-through site sitting in the link ends is connected back to back by short CAT 5E or other convenient cable as shown below and the daisy chained ports are configured as trunks

Link design and installation

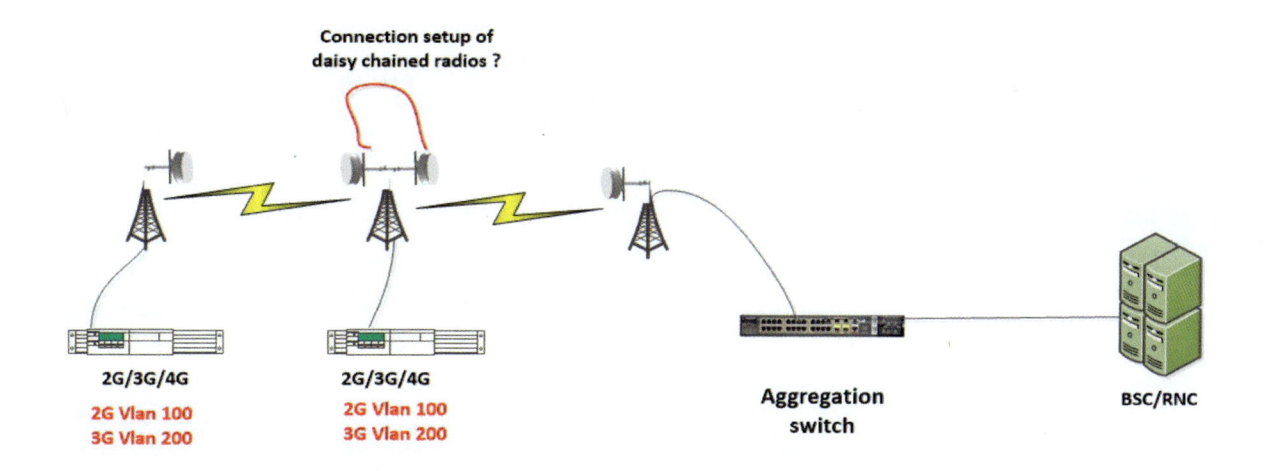

Figure 5.21 back to back radio traffic routing

Nowadays many microwave equipment providers came up with integrating external bulky router into the microwave transmission unit. In such case you will have microwave router in which L1 (radio), L2 (Ethernet), and L3 (IP) are integrated into a single chassis.

When microwave routers are deployed, routing should be configured for management traffic to pass through the daisy chained links of the backbone route. Open shorted path first (OSPF) is popular choice in which the links are grouped into areas. OSPF is dynamic routing protocol that is vendor-free

An example of routed configuration is shown below where OSPF is deployed

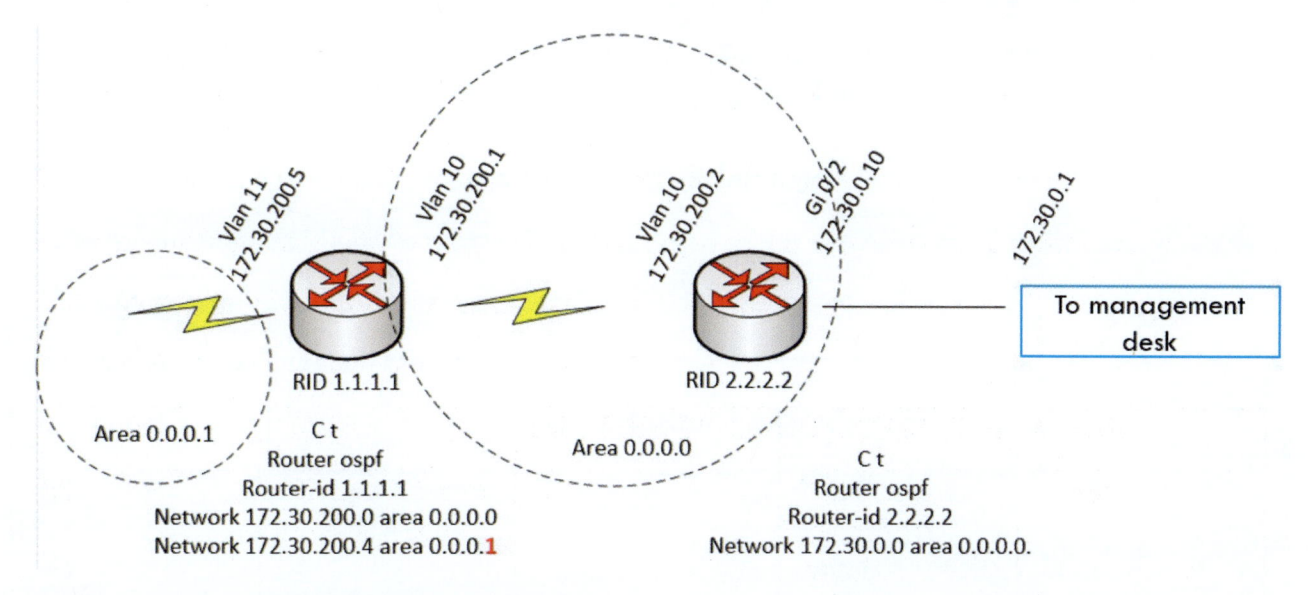

Figure 5.22 routed management for IP radio links

Each microwave is given router ID and areas are configured on the interfaces along with networks to advertise and exchange with its neighbors in the same area.

Link design and installation

Another important setup in interfacing microwave transmission links within themselves and with other network components is synchronization. Synchronous Ethernet (Sync-E) is popular standard that comes with all microwave transmission systems.

An example of synchronous Ethernet in typical microwave radio to synch with BTS connected to its g1 port is shown. This configuration is extracted from Aviat CTR radio

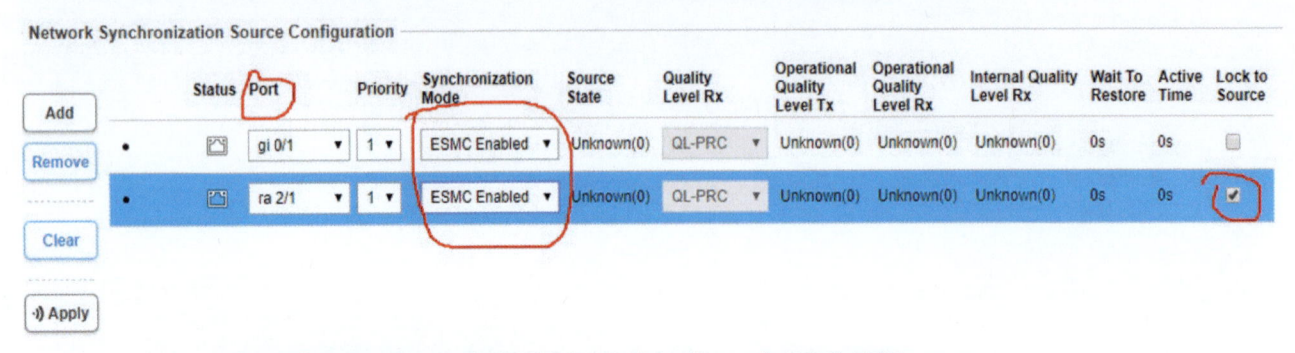

Figure 5.23 Sync E configuration example on Aviat CTR

A common mistake made by most engineers working on microwave radio switches is Ethernet loops created by multiple connection between two switches. This means if two cables are run between two adjacent switches and all those four ports connected by the two cables are enable, then loop will start which can bring the whole traffic passing in the switch. This is very hard to detect in early stages of troubleshooting.

Spanning tree protocol (STP) is another feature than comes with all microwaves Ethernet switches. This protocol will prevent loop by disabling port which has the lowest priority as shown below.

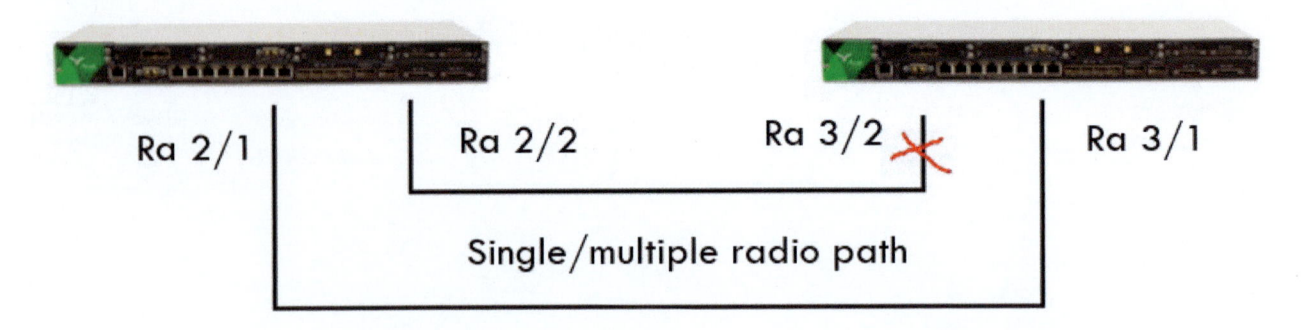

Figure 5.24 Spanning tree for Ethernet radio link loops

Normally network engineers find easy to troubleshoot routed or IP networks. They could use ping and traceroute IP utilities to locate location of network fault and reachability of remote sites. When it comes to Ethernet L2 networks they also got fault management utilities. When purchasing microwave radio equipment, ECFM (Ethernet connectivity fault management) is your choice. ECFM protocol (IEEE 802.1ag) comes with three utilities as shown below

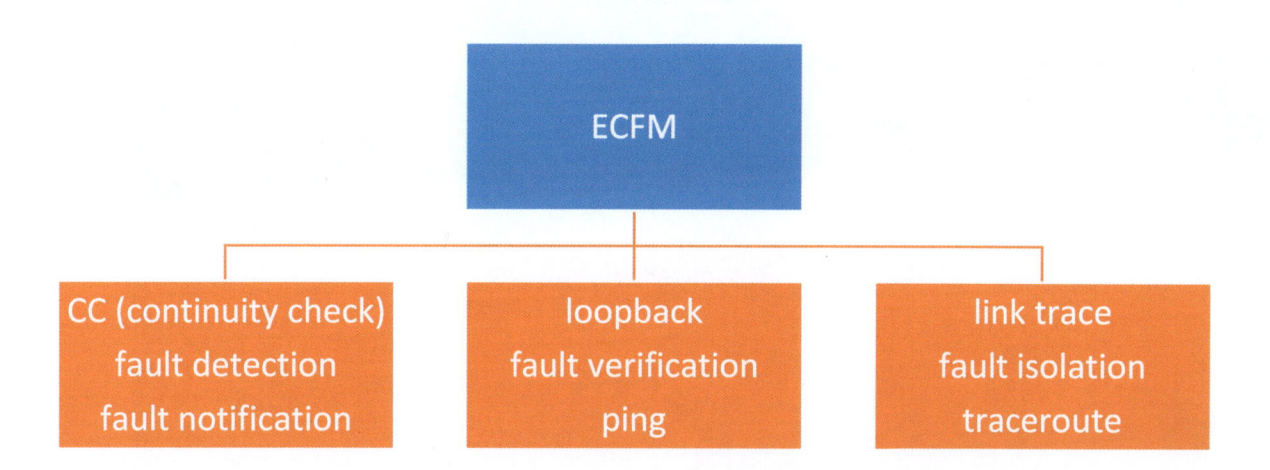

```
                    ┌──────────┐
                    │   ECFM   │
                    └────┬─────┘
         ┌───────────────┼───────────────┐
┌────────────────┐ ┌────────────────┐ ┌────────────────┐
│ CC (continuity │ │    loopback    │ │   link trace   │
│     check)     │ │                │ │                │
│ fault detection│ │fault verification│ │ fault isolation│
│fault notification│ │      ping      │ │   traceroute   │
└────────────────┘ └────────────────┘ └────────────────┘
```

Figure 5.25 ECFM Ethernet links provisioning

An example of Ethernet ping service is pinging remote MAC address to confirm reachability
#ping Ethernet Mac 00:e0:e2:96:2b:50 domain name vlan 10

```
aos# ping ethernet mac 00:e0:e2:96:2b:50 domain name vlan 10
Ping is initiated from first Possible MEP 200
Please use Interface Index and Direction to initiate from Specific MEP
Sending 1 Ethernet CFM loopback messages, timeout is 5 seconds
Success rate is 100.0 percent 1/1
```

Figure 5.26 ECFM ping example in Ethernet radio links

As shown above, ping succeeded with 100% success rate. And there is not fault from this node up to remote node.

Finally traceroute to remote MEP to trace end to end and intermediate maintenance points MAC addresses.
#traceroute Ethernet Mac 00:e0:e2:96:2b:50 domain name vlan 10

```
aos# traceroute ethernet mac 00:e0:e2:96:2b:50 domain name vlan 10
Trace is initiated from first Possible MEP 200
Please use Interface Index and Direction to initiate from Specific MEP

Traceroute to MAC address 00:e0:e2:96:2b:50 in domain name at level 5
with vlanId 10
------------------------------------------------------------------------
Hops    Host               Ingress MAC         Ingress Action    Relay Action

        Next Host          Egress MAC          Egress Action     Fwd Status
------------------------------------------------------------------------
1   00:e0:e2:96:0e:81:00:0d   00:e0:e2:96:25:d0    IngOK         RlyFDB
    00:e0:e2:96:25:c1:00:11   00:e0:e2:96:25:d2    EgrOK         Forwarded

2   00:e0:e2:96:25:c1:00:11   00:e0:e2:96:2b:50    IngOK         RlyHit
    00:e0:e2:96:2b:41:00:0f        NONE           EgrNoTlv       Terminal MEP

HOPS - 1  :
-----------
LTR Management Address: No management address was present in the LTR
Ingress PortId Subtype : 1
Ingress PortId         : radio2/1
Egress PortId Subtype  : 1
Egress PortId          : radio3/1

HOPS - 2  :
-----------
LTR Management Address: No management address was present in the LTR
Ingress PortId Subtype : 1
Ingress PortId         : radio2/1
Egress PortId Subtype  : NONE
```

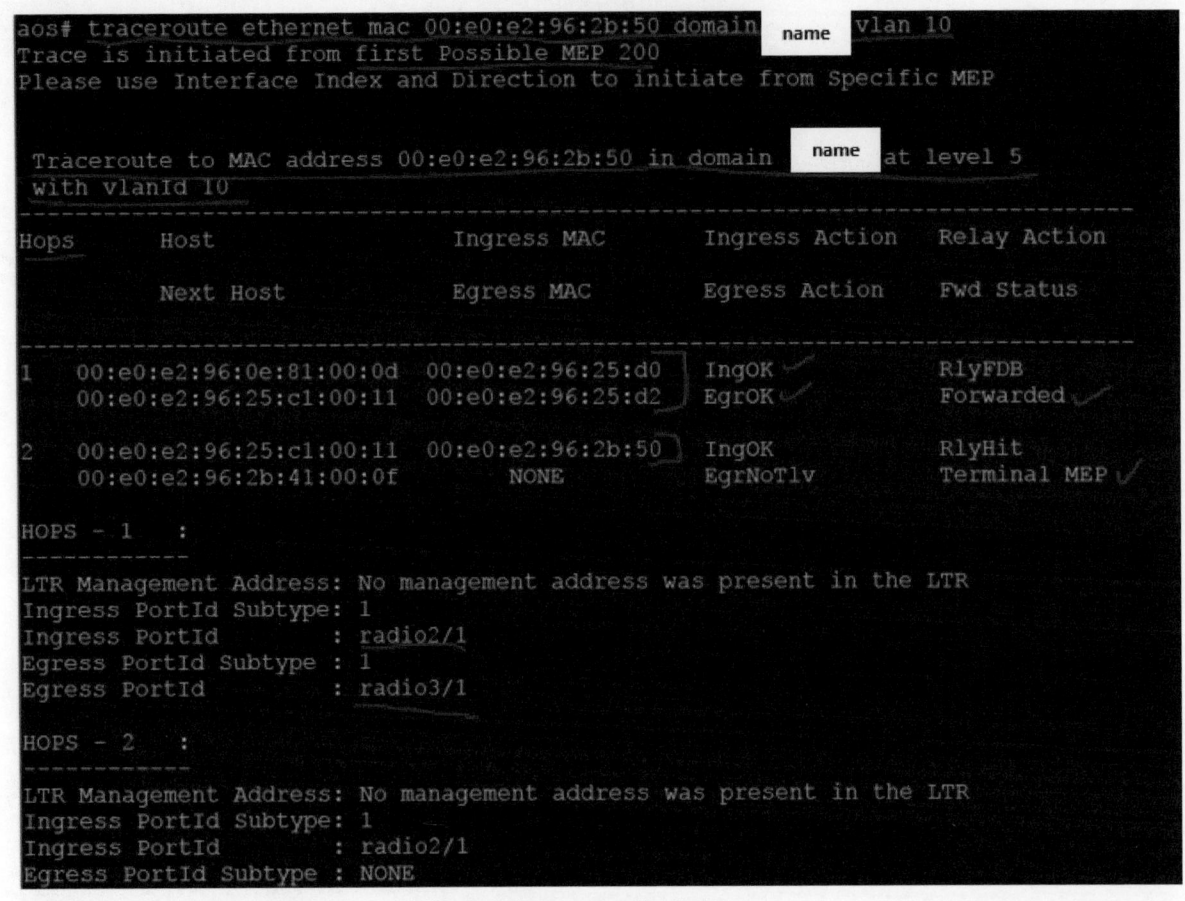

Figure 5.27 ECFM traceroute example in Ethernet radio links

The illustrations above are performed on Aviat CTR router by the author but most other microwave vendors support ECFM protocol in microwave L2 devices

We conclude this chapter by looking at ring protection used to protect operator network but often not used. When radio links are connected in ring fashion, protection as well as loop protection will be performed by licensing and implementing ERPS (Ethernet ring protection switching) feature. ERPS operates as follows:

- Ethernet ring protection switching (G.8032) → 50ms protection switching
- RPL (Ring protection link) is blocked under no failure condition
- RPL link lies between RPL owner and RPL neighbor.
- When a link in the ring fails, signal failure (SF) message is sent across the ring. When the RPL owner detects SF, it unblocks its port.
- When ring node receives NR, RB messages, and fault is cleared, it unblocks blocked port.
- Uses ring protection switching PDUs (R-APS) to communicate among the nodes; SF, NR, RB

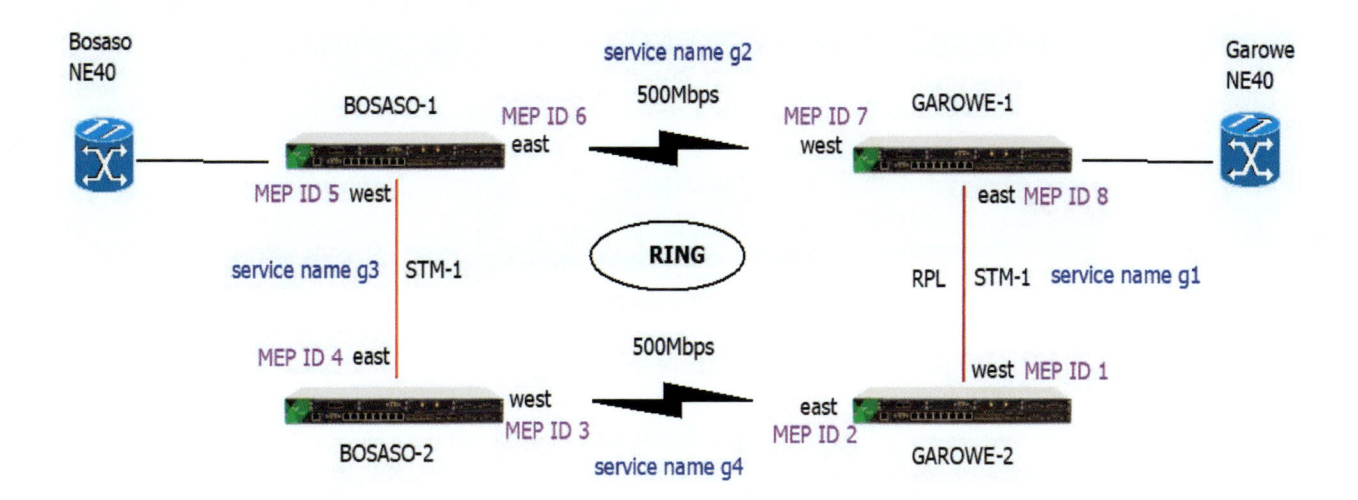

Figure 5.28 ERPS case example in Ethernet radio rings

ERPS is good protection switching when the radio or fiber links are L2 and can only transport Ethernet traffic

Competent transmission engineer's notes on wireless capacity and outages

Wireless networks are good choice when you are planning easy design and deployment. But they are prone to topographical and environmental effects. Moreover the wireless spectrum is shared and bandwidth is limited by system noise and regulations.

Capacity of the wireless channel is improved if we use wide-band spectrum (at 70 GHz). However, more bandwidth means more processing power and complex circuits. More bandwidth also expensive. Capacity can also be increased at the physical layer by aggregating multiple channels into one, such as when the vertical and the horizontal polarization of t

Also at the Ethernet frame level, we can compress more data frames and put them into smaller channel (more data bits per hertz). Capacity can also be increased by removing redundancy bits to increase the size of the payload.

On the other hand, frequent errors will appear in the wireless network if ill-planned, or installed in unprofessional way. Some of the outages include

- Multi-path fading which results from reflectors on the way of the signal such as tall buildings, trees and moving objects. The signal then travels in many paths with different speed and phase. At the receiver the signals arrive and cause destructive interference.
- Noise which adds to the bandwidth of the wanted signal. The noise may come from the environment, electric lines, motors and generators, air conditioners, rusted bolts, or it can come from the system itself. There is antenna noise, feeder noise, wave-guide noise, free space path loss, and so on
- Interference can also result bad performance if two nearby transmitters use same frequency (co-channel) or neighboring frequency (adjacency-channel).

Ways of mitigating wireless network outages

- Correct site survey before deploying any wireless network
- Detailed link analysis and optimum path selection in case of point to point LOS links
- Get equipment with low losses
- Deploy redundant power infrastructure or uninterruptible power supply to avoid critical site shutdown
- Carry out spectrum analyzer field testing to locate possible sources of interference.
- Employ professional installers when deploying new site.
- Ground all devices correctly to avoid lightning surges.

- Site inspection and troubleshooting should be done by experienced personnel to avoid human error
- Perfect alignment of point to point antennas is also very important.
- Seat well and tighten all cables and connectors. do this regularly
- Prepare spares at all remote site, to reduce troubleshooting time
- Create log file to record alarms and troubleshooting techniques. This will help if similar problem happens again.
- Develop network management techniques, such as pro-active problem solving and not reactive problem solving
- Employ space diversity for path redundancy and hot-standby for equipment protection.
- Use adaptive equalizers to smooth receiver filter response.
- Devise preventive maintenance plan. There is nothing worse if you do not maintain your network and leave smaller issues grow into bigger one that can bring the whole network down.

Typical fault scenarios seen in microwave radio communication is summarized in the table below. These can disrupt customer traffic and satisfaction

Table 5.9 examples of radio link performance issues

Alarm	Possible cause	Solution
No communication management with remote terminal

Remote communication failure | Pathloss and equipment malfunction | - If you see other traffic affecting path alarms, suspect path failure
- If you don't find traffic affecting alarm, then suspect equipment failure
- Check for other equipment related alarms, and narrow problem to specific module.
- Check if communication failure if one-way or two-way
- One-way due to RF-processing stages
- Incorrect IP addressing set for one of the |

		terminals
Radio path down	Path loss or equipment malfunction Pathloss caused by rain (>11GHz), diffraction/multipath(<11GHz Check for path loss(both-way, low RSL, BER alarms) Check equipment (ODU, modem)	• Traffic will be affected in both directions • Complete pathloss preceded by RSL and BER alarms • Equipment related path problem indicated by hardware or software alarms at local or remote
Silent TX failure (remote)	Both remote online IDUs are in receive alarm (path failure). TX mute applied on the online TX of HS or SD protected pair. No detected remote receive signal (demodulator unlocked).	• Turn on remote TX • Replace ODU • Remote site power failure
Configuration not supported	Incorrect node plug-in installed License violation	• Install correct plug-in • Apply correct configuration • Check the license card
BER threshold exceeded	Propagation due to rain and multipath fading Equipment malfunction Interference	• Check RSL at both ends of the link • If RSL is abnormal on one end, scan the area for interference

In many cases (especially where redundancy power infrastructure is not available) power failure is the most common cause of link path failure. An example witnessed by the author and solved is shown below where the -48V from the rectifier was lost and the whole remote site shut down

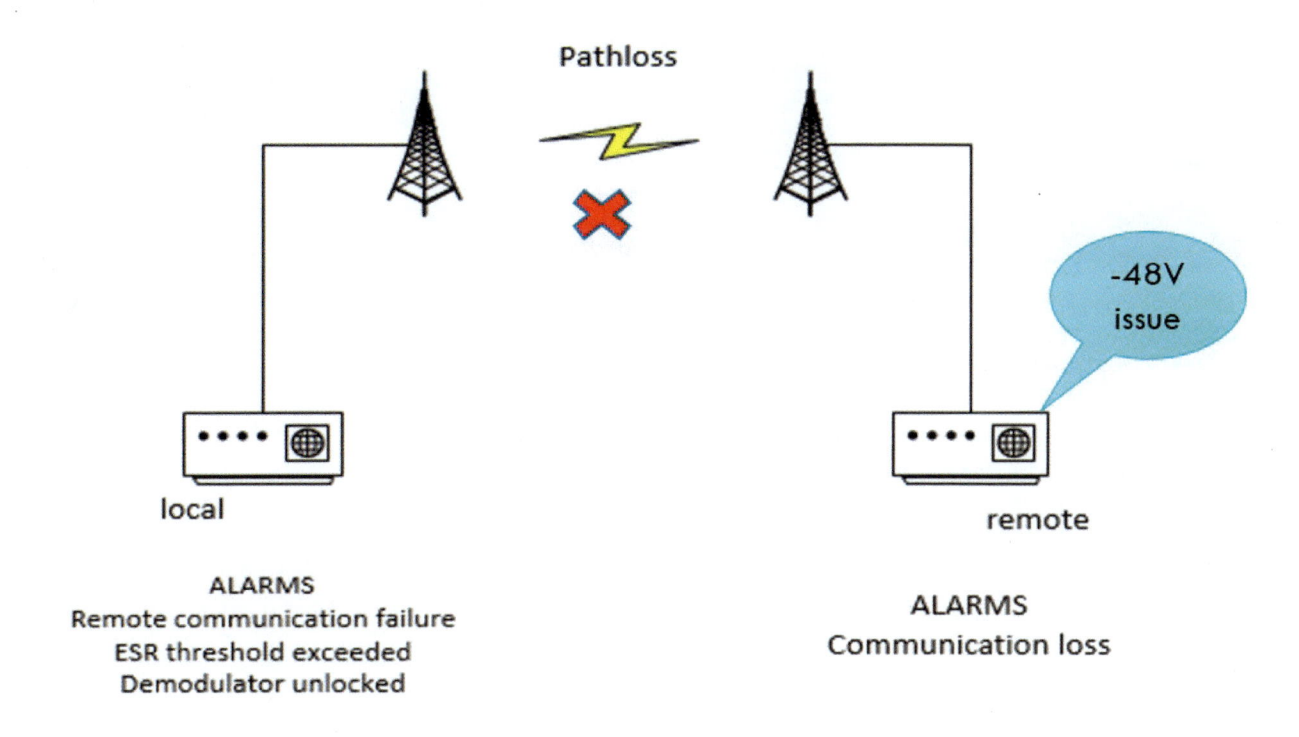

Figure 5.29 illustration of radio path failure alarms

Networking and protocols

The best place to start when talking about is OSI model which has seven layers as shown in table 5.10 below

Table 5.10 OSI model

	Layer	Function	Example of protocols
7	Application	Interface to user	HTTP/FTP
6	Presentation	Formatting and encryption	SSL
5	Session	Session for separate traffic	PPTP
4	Transport	Data segmentation	TCP/UDP
3	Network	Logical IP addressing	IPv4/IPv6
2	Data link	Physical MAC addressing	Ethernet
1	Physical	Bits for transmitting over the medium (e.g. microwave)	Microwave

An engineering working on ICT networks should have basic understand on network components and how to use each of them. Some network components are shown in figure 5.30 below

Link design and installation

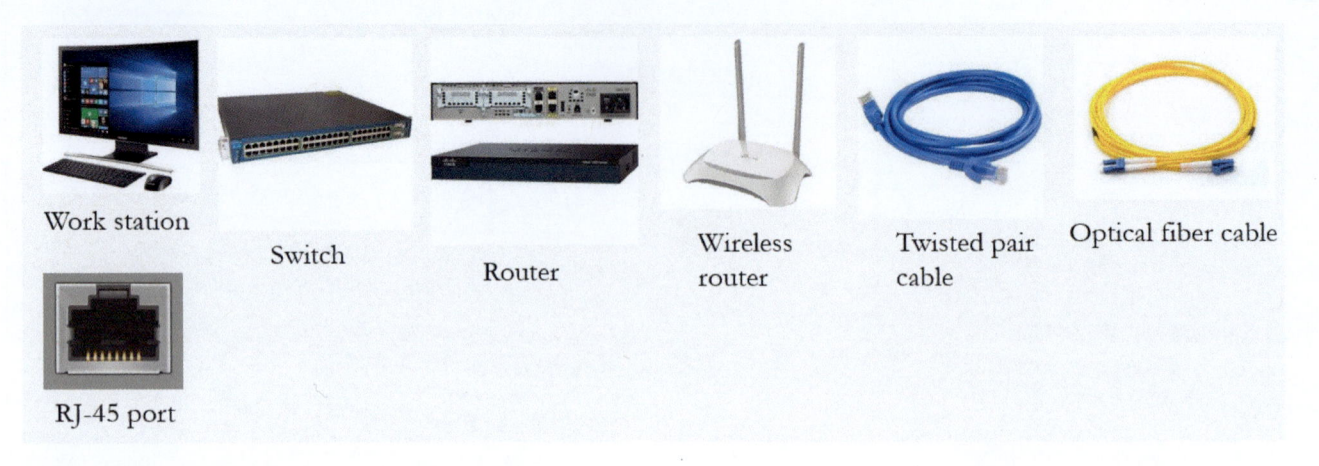

Figure 5.30 network components

To provide better user experience with high system availability redundancy is a must option that comes with high CAPEX and OPEX. Redundancy occurs at various stages of the overall network as shown in figure 5.31 below

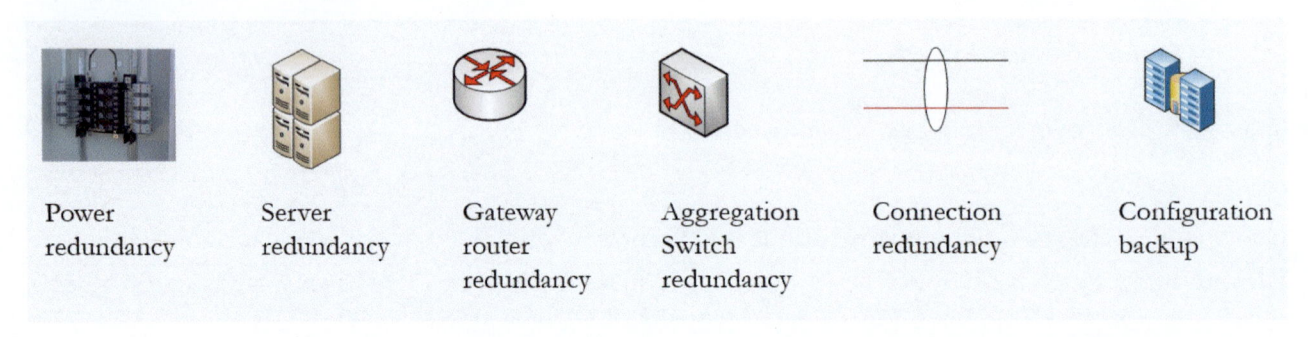

Figure 5.31 network redundancy aspects

Another important consideration in the network planning process is also designing the network by understanding design requirements. A typical network design requirement is shown in figure 5.32 below

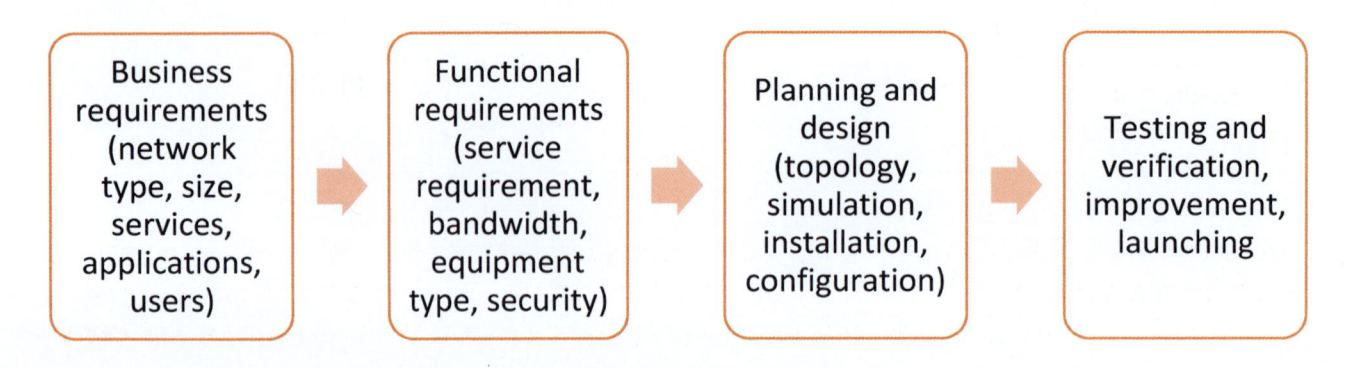

Business requirements (network type, size, services, applications, users) ➡ Functional requirements (service requirement, bandwidth, equipment type, security) ➡ Planning and design (topology, simulation, installation, configuration) ➡ Testing and verification, improvement, launching

Figure 5.32 network design requirements

An engineer working on microwave transmission system should also have basic knowledge and experience on Ethernet switching and IP routing. Ethernet switching include how to configure VLANs, STP, Ethernet over WAN such as MPLS L2 VPN. IP routing include RIP, OSPF, BGP, MPLS L3 VPN, and IPv6

Ethernet switching links are hard to troubleshoot than the IP. Some of the troubleshooting tips for Ethernet links are:

- Verify VLAN configuration
- Check switch ports are enabled
- Check possible loops and enable STP
- Check MAC-address-table for learned devices
- Check port status, MTU and duplex
- Check interface counters to monitor traffic

In the past transmission links were using PDH and SDH to carry radio frames. In the system room, many racks were placed that provided termination of E1 lines carrying voice and small data services. E1 circuits were very efficient, reliable and easy to troubleshoot. They are still in use today in GSM networks

Now Ethernet and IP services were run in LAN where RJ45 wired cables were used. Ethernet over wireless then emerged which made microwave transmission links to support IP traffic such as that of 3G and 4G. These presents challenges like Ethernet loops and IP conflicts. But with good planning and vast technical training, these challenges can be solved

In this section we will talk about Ethernet and IP transmission WAN connections

- Metro Ethernet
- MPLS L2 VPN
- MPLS L3 VPN

Metro Ethernet is an extension of local area wireless networks to microwave long distance links. This means the whole microwave route of the radio links can be pictured as one big Ethernet switch where users located at different geographical points connect to its ports as shown below

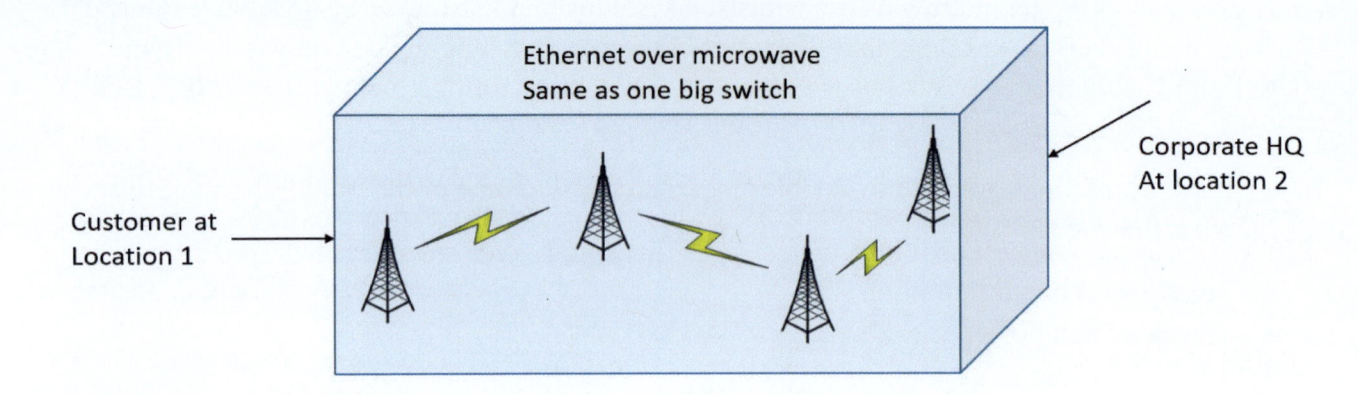

Figure 5.33 Ethernet over microwave radio

In Metro Ethernet environment, VLANs are configured on the microwave switches to carry the traffic

Many operators now seem to be interested IP microwave to replace Ethernet. This will require a router to replace the switch. The current deployment mixed Metro-Ethernet and IP MPLS. The advantage of IP links is that MPLS VPN can be used for private business connections

IP links work on static and dynamic routing protocols to forward IP packet on the network. IP routing is similar to sending mail through the mail carrier offices. The senders sends the mail inside an addressed envelope to his local mail office. The local office will then determine the shortest route to the next mail office until the mail reaches the receiver. This comparison between mail delivery and IP packet delivery is shown in figure 5.34

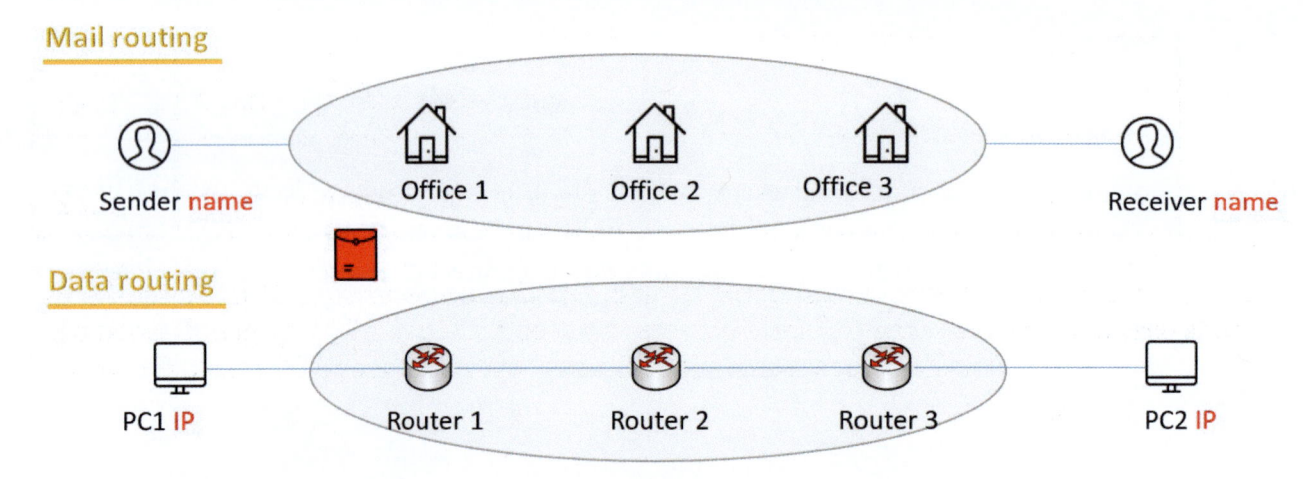

Figure 5.34 IP routing conceptual

Routing protocols forward IP traffic from one router (hop) to the next neighboring router (hop) using the best path between source and destination

Two types of routing protocols

- Static routing in which the ICT engineer manually configures the best path

- Dynamic routing protocols in which the routers dynamically determines the best path

Routers keep routing database into routing tables which will contain all connected and learned networks through static and dynamic protocols configured on the routers

```
       192.168.50.0/24 is variably subnetted, 5 subnets, 4 masks
S      192.168.50.0/26 [1/0] via 192.168.50.97
C      192.168.50.96/27 is directly connected, GigabitEthernet0/0
L      192.168.50.98/32 is directly connected, GigabitEthernet0/0
C      192.168.50.144/28 is directly connected, GigabitEthernet0/1
L      192.168.50.145/32 is directly connected, GigabitEthernet0/1

O      172.16.50.12/30 [110/2] via 172.16.50.17
```

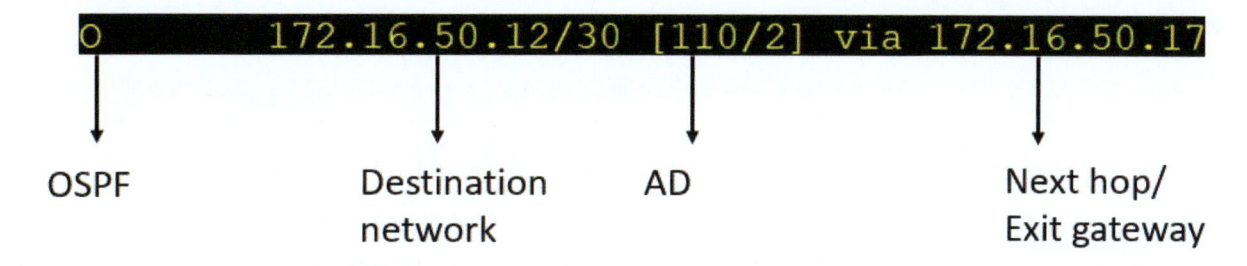

OSPF — Destination network — AD — Next hop/ Exit gateway

Figure 5.35 IP routing table content

The first thing the router places in its routing table is directly connected networks to its interfaces

When different routing protocols are configured on the same router, the best protocol to use to a given destination is decided by AD (administrative distance). It is a number which tells the best route to take when we have different routing protocols → lowest AD wins

For example the network in figure 5.36, the static route is chosen as the best route and added to the routing table because of its smaller AD compared with the OSPF

Routing protocol	AD
Directly connected	0
Static route	1
OSPF	110

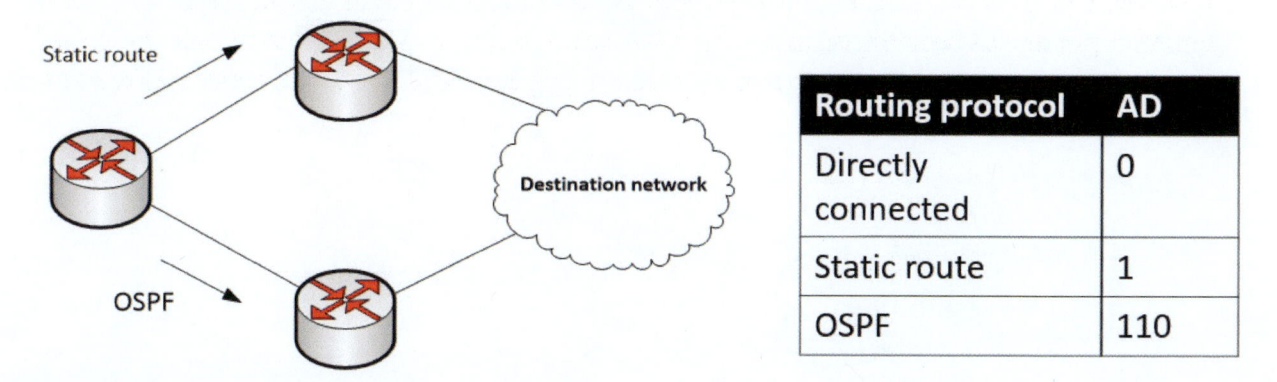

Figure 5.36 IP routing administrative distance (AD)

An example of static route configuration when a WAN links connects two company branches using different subnets is illustrated in figure 5.37.

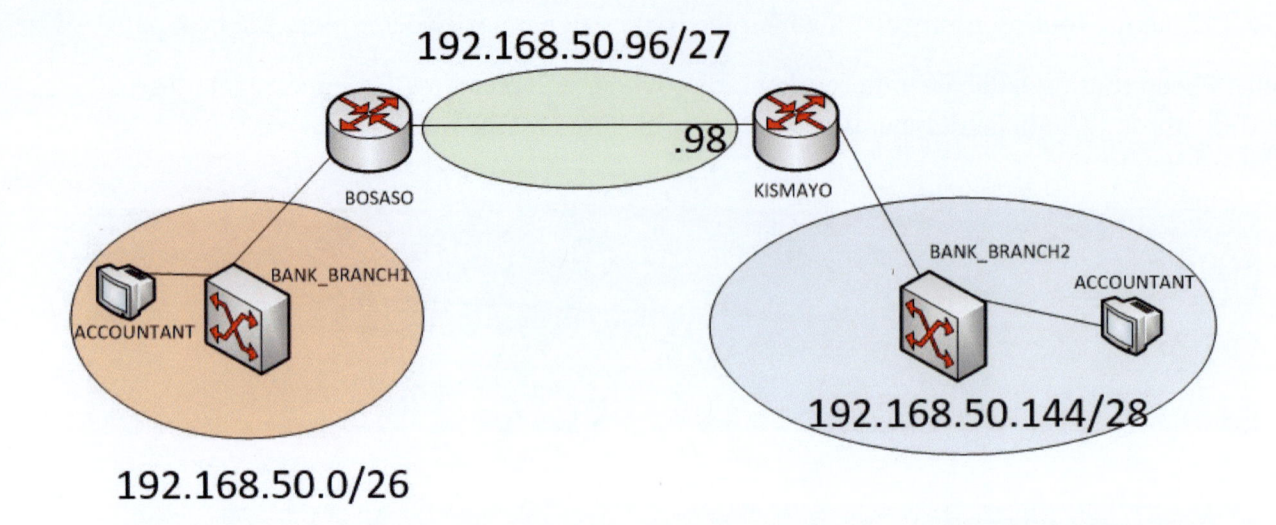

Ip route [destination network address] [destination subnet mask] [next hop]

Ip route 192.168.50.144 255.255.255.240 192.168.50.98 for BOSASO

Figure 5.37 Static routing on IP links

A special case of static route is called default route and used when the router does not know the best route towards a destination. It then forwards the traffic to the default route. It is used when an operator connects to an internet service provider (ISP). All traffic going to the internet is sent to the ISP gateway by configure one single default route. The following figure may help clarify this point

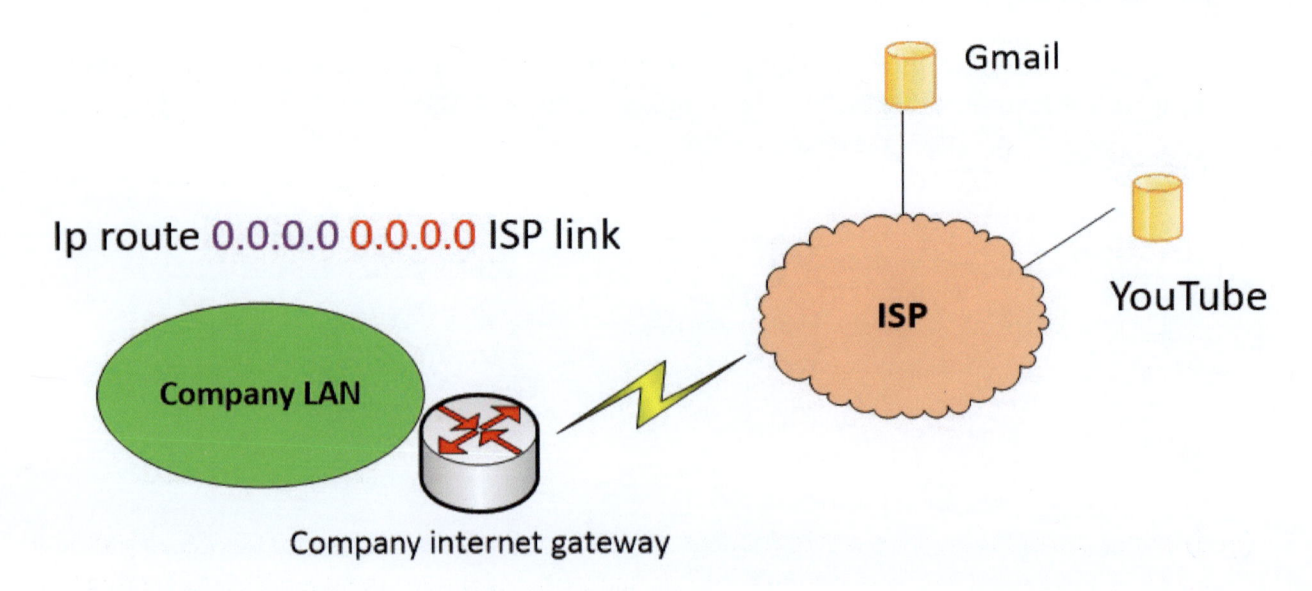

Figure 5.38 IP default route

We finalize L3 IP links with troubleshooting tips as presented in table 5.11 below

Link design and installation

Table 5.11 troubleshooting L3 networks

Issue	Solution
Destination host unreachable	Check routing table if destination network is missing Check PC default gateway
No internet connection	Check default route to the ISP
Request timeout	Check Firewall blocking return traffic
Other IP related issues	Check interface IP configuration and status

Next we talk about high speed WAN connection called MPLS

MPLS or multi-protocol label switch labels IP packets as they enter the provider network. This means VPN or virtual private networks can be deployed on MPLS infrastructure to create separate tunnel for each customer over single shared microwave radio. In the early phase of operator project management, decision has to be made whether to add this MPLS VPN into the licenses BOQ for equipment ordering. Operators who want to provide IP cloud computing and private links need this feature. Also internet service providers connecting inter-continental fiber optic links to the country need this feature as they provide separate internet to mobile network operators

An example of MPLS VPN over microwave radio routers is illustrated in figure 5.39 below

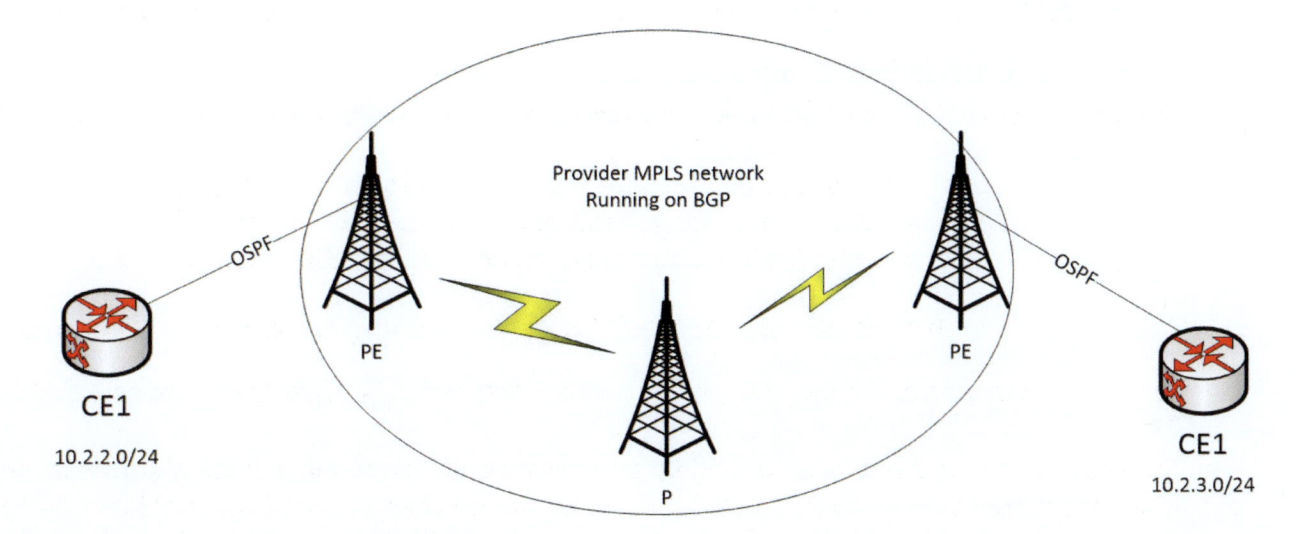

Figure 5.39 MPLS L3 VPN over microwave

Typically it is always good for the customer to opt for L3 VPN, as configuring the connection between the customer router and PE will be operator's responsibility. The idea of VPN over the MPLS routers is that different routing protocols could be configured for each connection. The component of MPLS L3 VPN are

- Customer edge (CE) router that is located on the customer premises and connects to PE with normal IP link
- Provider edge (PE) router that labels the IP packets from the customer. It could be microwave radio with an integrated router. These routers also configures VRF (virtual routing and forwarding) to separate different routing tables from different CE router on the VPN network. PE routers in the provider MPLS cloud connect with BGP protocol and has two interfaces. IP interface to CE and MPLS interface to P

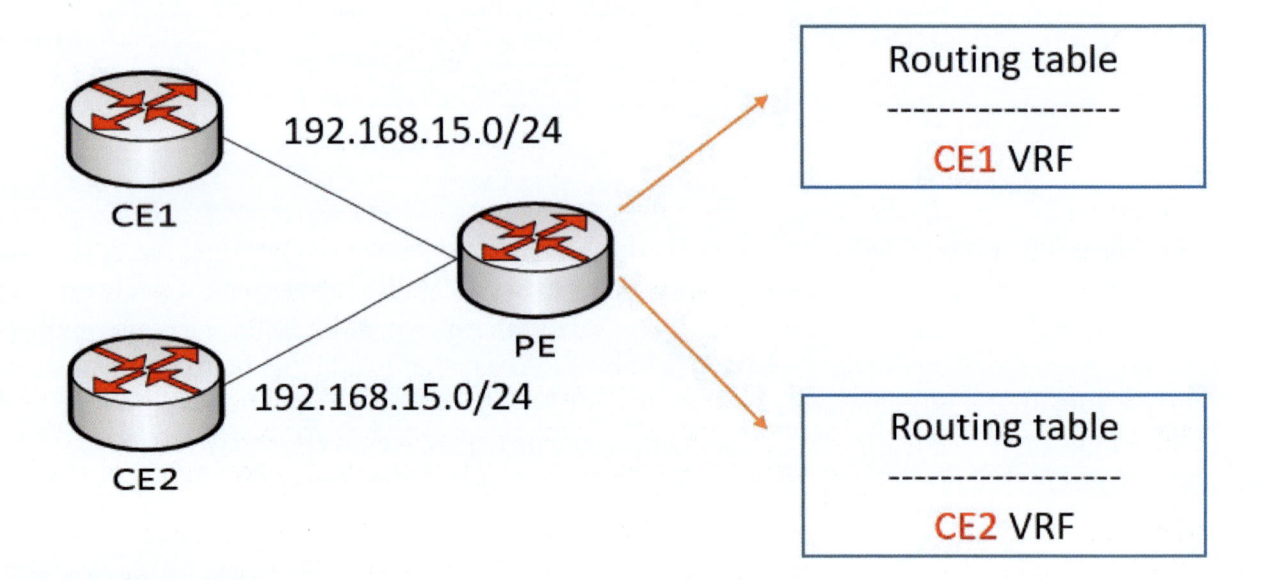

Figure 5.40 customer VRF

- P (Provider router) in the core network of the WAN operator and will switch the MPLS label across the MPLS core network. An example of MPLS label captured with Wireshark packet sniffer is shown below

Link design and installation

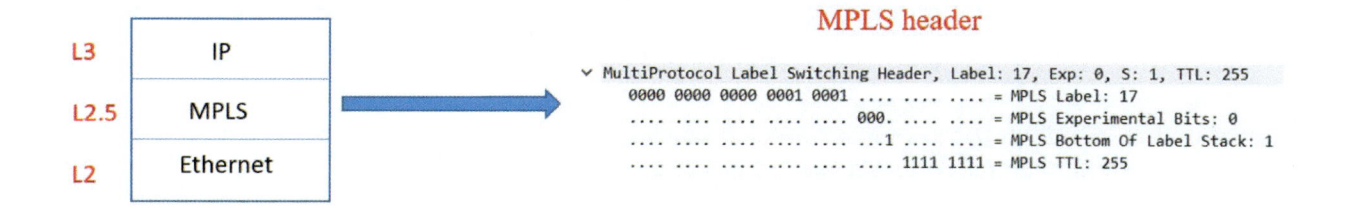

```
L3    | IP       |
L2.5  | MPLS     |   →   MPLS header
L2    | Ethernet |
```

```
∨ MultiProtocol Label Switching Header, Label: 17, Exp: 0, S: 1, TTL: 255
    0000 0000 0000 0001 0001 .... .... .... = MPLS Label: 17
    .... .... .... .... .... 000. .... .... = MPLS Experimental Bits: 0
    .... .... .... .... .... ...1 .... .... = MPLS Bottom Of Label Stack: 1
    .... .... .... .... .... .... 1111 1111 = MPLS TTL: 255
```

MPLS label is used between PE and P router, it is not used between PE and CE link

MPLS label is unidirectional ➔ different label used for forward and return traffic

Figure 5.41 MPLS header

This is pretty fast network as each P router in the core network does not need to read the routing table to determine the next hop, it only switches the label to the next P router

How does MPLS speaking routers establish relationship? They use LDP (label distribution protocol) in the core MPLS network. LDP routers listen TCP port 646 to monitor LDP sessions.

The following diagram shows the steps taken when configuring a typical MPLS L3 VPN connection in which OSPF is used for CE and PE link

```
[Configure MPLS on core routers on the provider network] → [Configure customer VRF on PE routers] → [Configure OSPF instance for each customer]
                                                                                                              ↓
[Hide core MPLS from customers to prevent customer route injection] ← [Redistribute OSPF and BGP so that end-to-end L3 VPN works] ← [Configure iBGP on PE routers]
         ↓
[Verification and testing]
```

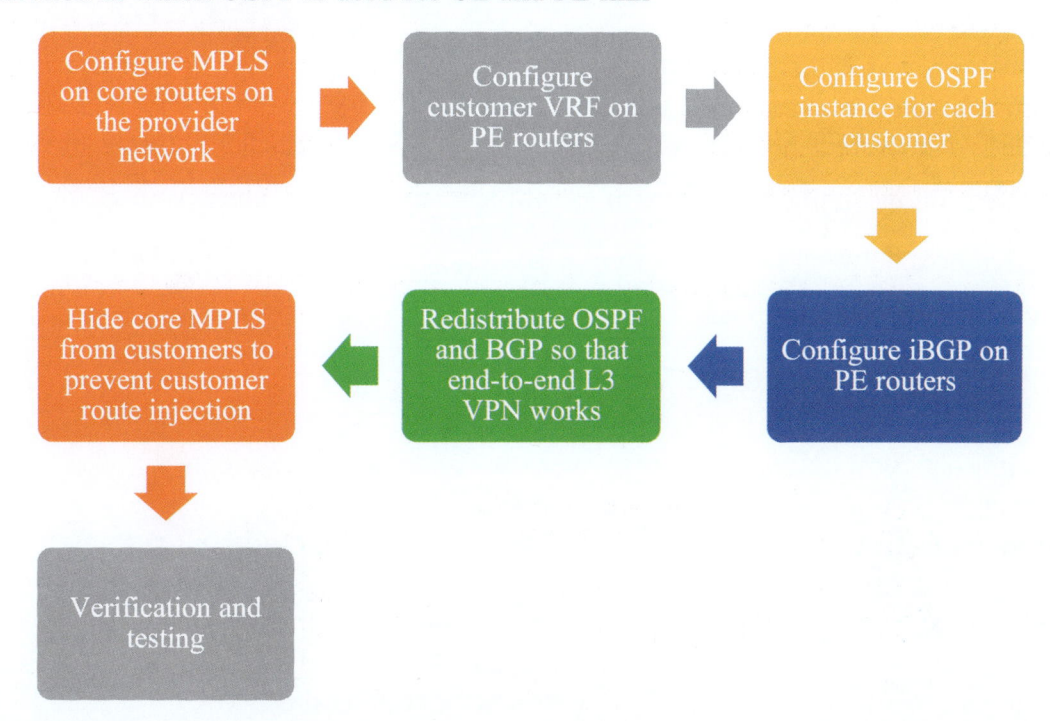

Figure 5.42 MPLS L3 VPN configuration steps

BGP or border gateway protocol is wide area network routing protocol used by internet service providers (ISPs) on the internet. It is an exterior gateway routing protocol (EGP) as opposed to interior gateway protocol (IGP) such as OSPF in the sense that BGP connects different autonomous system on the internet. Each ISP on the internet will be assigned public autonomous system (AS) number by IANA (internet assigned numbers authority). This same organization is also responsible for assigned public IP addresses to ISPs in which internet registers (IR) at regional and national levels take part in the assignment. As an example figure 5.43 below shows a typical public IP addresses assignment to telecom operator in Africa

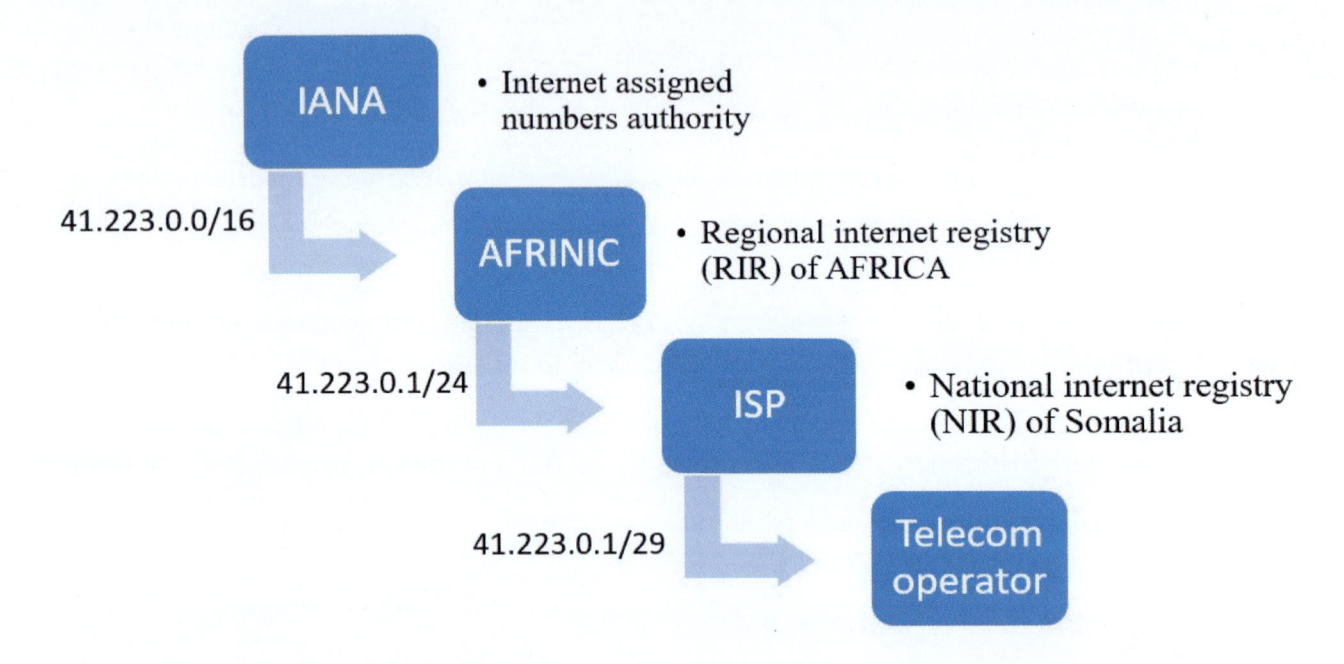

Figure 5.43 Public IPv4 assignment process

Public AS numbers follow similar procedure of assignment and have the range 1 - 64511

Some of the differences between OSPF and BGP is summarized in table 5.12 below

Table 5.12 OSPF and BGP comparison

OSPF	BGP
Neighbors dynamically form between routers using hello message	Neighbors configured explicitly
Within an autonomous system	Between different autonomous systems
Hello multicast message	Uses TCP protocol port 179
Link state routing protocol	Path-vector routing protocol
Best path selection based on metric	Best path selection based on path attributes (PA)

BGP can be interior BGP (iBGP) used within a single autonomous system or exterior BGP (eBGP) used between two different autonomous systems. Thus eBGP is normally the connection between the telecom operators towards one or more internet ISP. An example of iBGP and eBGP configuration using Cisco routers is shown in figure 5.44. The operator has AS of 10 and has two internet connection using BGP. One to ISP1 with AS 1 and another to ISP2 with AS 2. The connection between the operator and the ISP uses eBGP while iBGP is used within the telecom operator routers (the grey circle network). The configuration the internet facing router of the operator, G1 is illustrated in the boxes. The left box is eBGP towards ISP1 and the right box is iBGP towards G2

```
G1(config)#router bgp asn 10
G1(config-router)#neighbor 1.1.1.1 remote-asn 1 (ISP1)
```

```
G1(config)#router bgp asn 10
G1(config-router)#neighbor 2.2.2.2 remote-asn 10 (G2)
```

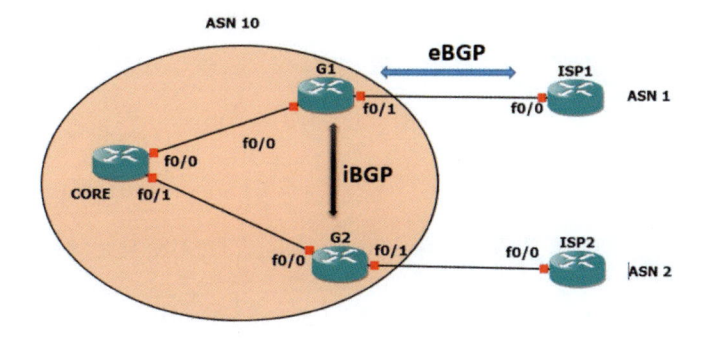

Figure 5.44 iBGP and eBGP configuration

IGP protocols such as OSPF use metric to determine best path. OSPF uses metric based on link bandwidth for example. BGP uses not just a metric but uses a number of path attributes (PA) to determine best route towards destination network on the internet. A summary of different PA used by BGP is in table 5.13

Table 5.13 BGP path attributes

Next hop	How many hops the prefix is away?
AS Path	How many ASNs the prefix is away?
Local preference	Used to influence best outbound route for all routers inside ASN
Origin	Routes injected from IGP
Multi-exit discriminator (MED)	Routers in different ASNs can influence in terms of BGP decisions

The AS PATH attribute is based on selecting the route with fewer AS count. Figure 9.45 illustrates BGP route advertisement using AS PATH attribute

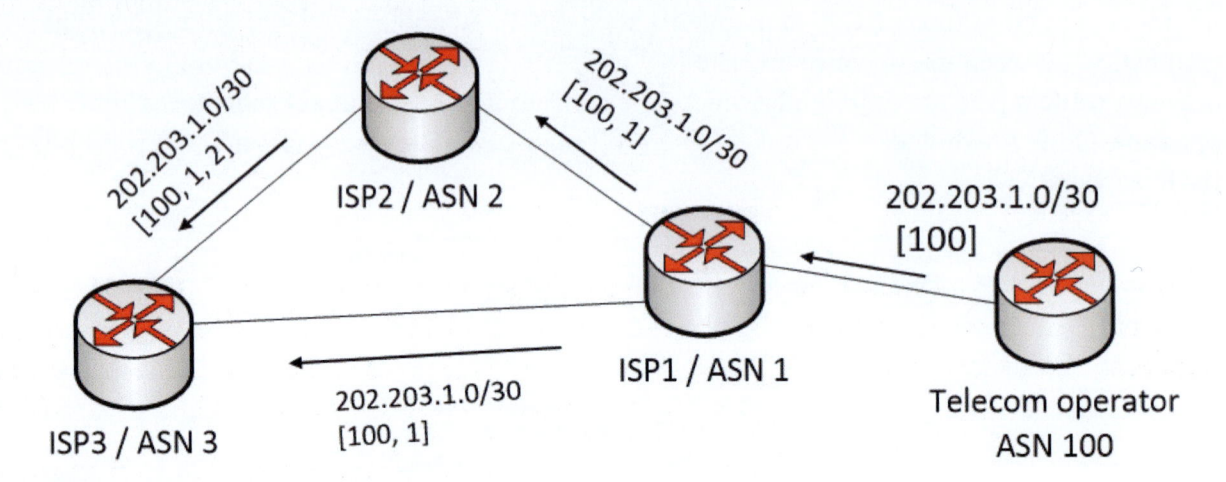

Figure 9.45 AS PATH advertisement

ISP3 learns two route for the prefix 202.203.1.0/30 advertised by the telecom operator

It add the lower route to its BGP table as best path because it has small number of ASN [100, 1]

When an operator has a single outbound route (towards the internet), BGP may not be necessary and static route will be sufficient. A default route towards the internet ISP and static route towards the public IP behind the operator gateway. An example is shown in figure 9.46

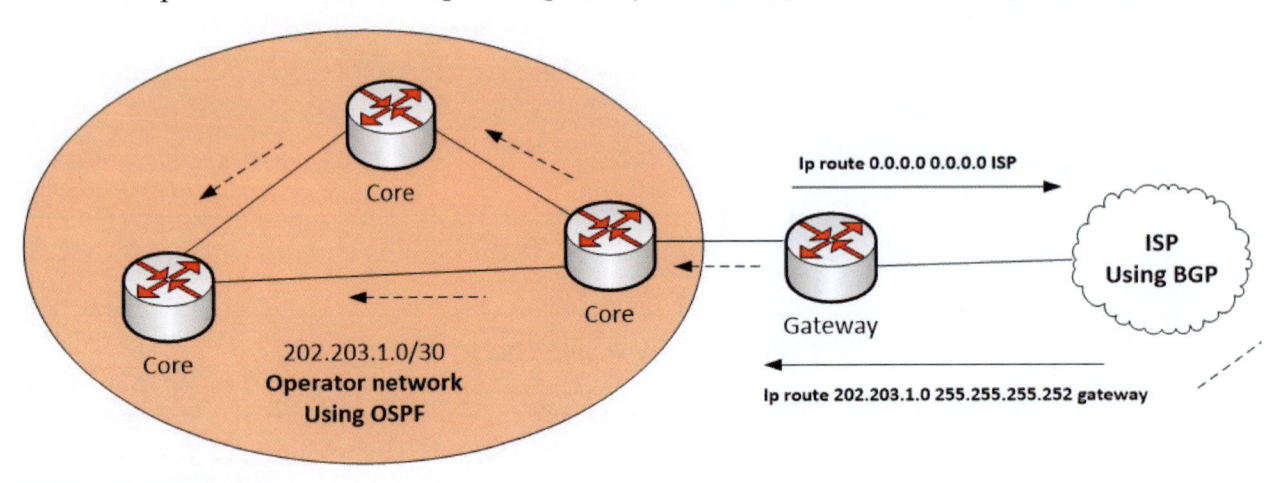

Figure 9.46 default route pointing to the internet

The default will then be redistributed into the operator OSPF domain

BGP becomes more important in cases two or more outbound routes to internet exist and one route is to be preferred over another for specific internet destinations

The example below is an enterprise learning internet default route from two internet providers, ISP1 and ISP2, along with ISP1 router configuration

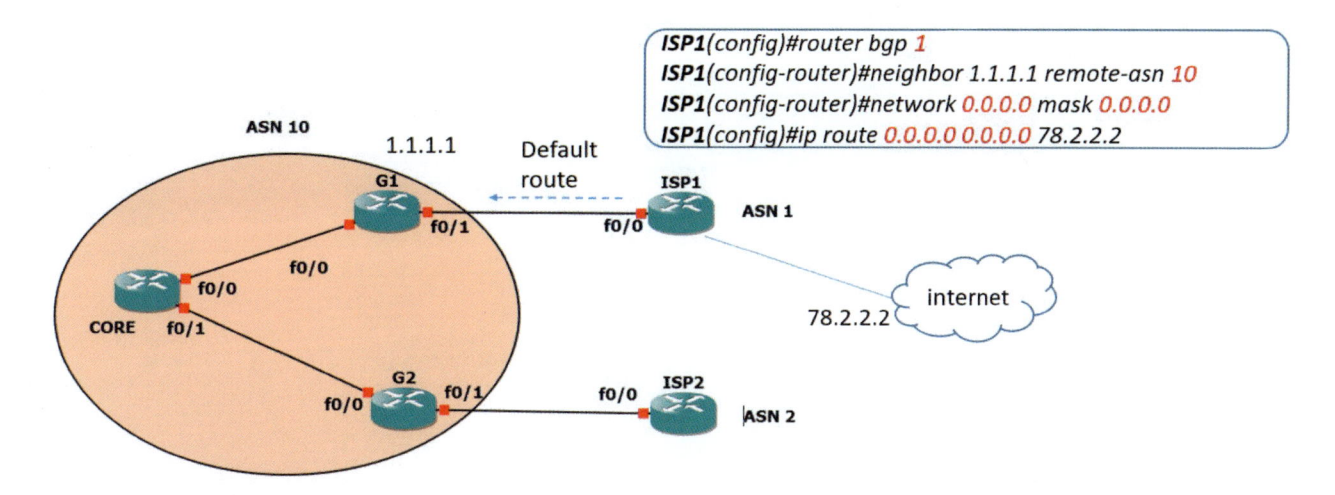

```
ISP1(config)#router bgp 1
ISP1(config-router)#neighbor 1.1.1.1 remote-asn 10
ISP1(config-router)#network 0.0.0.0 mask 0.0.0.0
ISP1(config)#ip route 0.0.0.0 0.0.0.0 78.2.2.2
```

Figure 9.47 ISP advertising default route through BGP

The snapshot below shows G1 has learnt default route through BGP

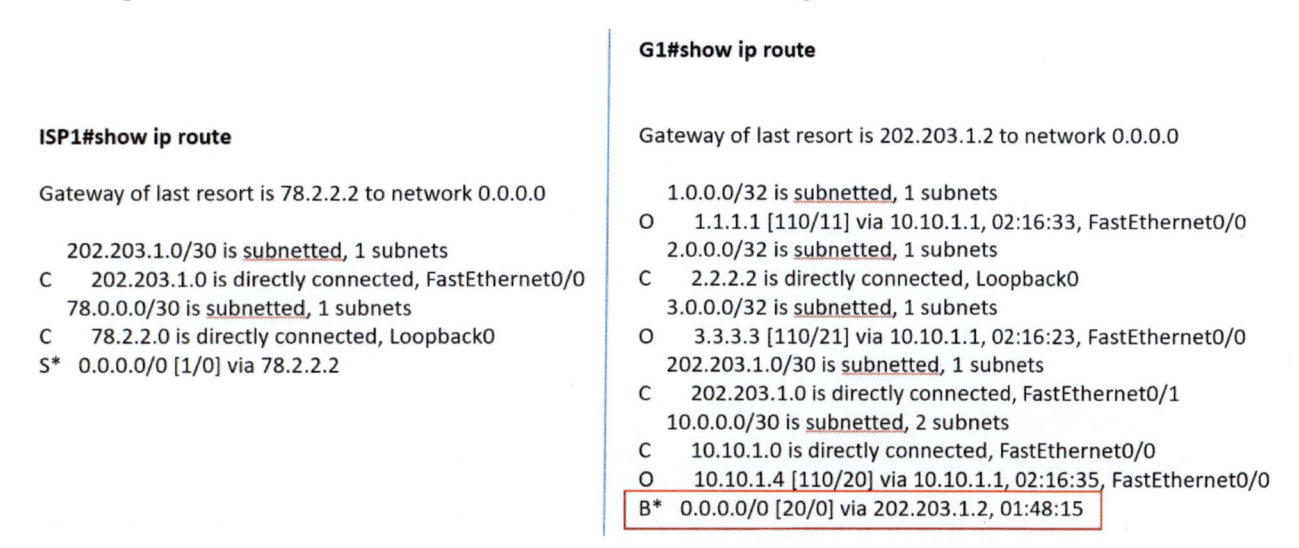

ISP1#show ip route

Gateway of last resort is 78.2.2.2 to network 0.0.0.0

    202.203.1.0/30 is subnetted, 1 subnets
C   202.203.1.0 is directly connected, FastEthernet0/0
    78.0.0.0/30 is subnetted, 1 subnets
C   78.2.2.0 is directly connected, Loopback0
S* 0.0.0.0/0 [1/0] via 78.2.2.2

G1#show ip route

Gateway of last resort is 202.203.1.2 to network 0.0.0.0

    1.0.0.0/32 is subnetted, 1 subnets
O   1.1.1.1 [110/11] via 10.10.1.1, 02:16:33, FastEthernet0/0
    2.0.0.0/32 is subnetted, 1 subnets
C   2.2.2.2 is directly connected, Loopback0
    3.0.0.0/32 is subnetted, 1 subnets
O   3.3.3.3 [110/21] via 10.10.1.1, 02:16:23, FastEthernet0/0
    202.203.1.0/30 is subnetted, 1 subnets
C   202.203.1.0 is directly connected, FastEthernet0/1
    10.0.0.0/30 is subnetted, 2 subnets
C   10.10.1.0 is directly connected, FastEthernet0/0
O   10.10.1.4 [110/20] via 10.10.1.1, 02:16:35, FastEthernet0/0
B* 0.0.0.0/0 [20/0] via 202.203.1.2, 01:48:15

Figure 9.48 operator learns advertised default route through BGP

The operator is also required to advertise its public IP addresses to the internet. Normally this network runs on OSPF and router G1 will redistribute this OSPF into the BGP domain. An example of the operator advertising its public prefix towards the internet is shown 9.49 below. Notice that a route-map is first created to advertise only the public prefix and not private address within the organization (the 10.10.1.0 is private address though, it is used to simulate public prefix as an example)

Link design and installation

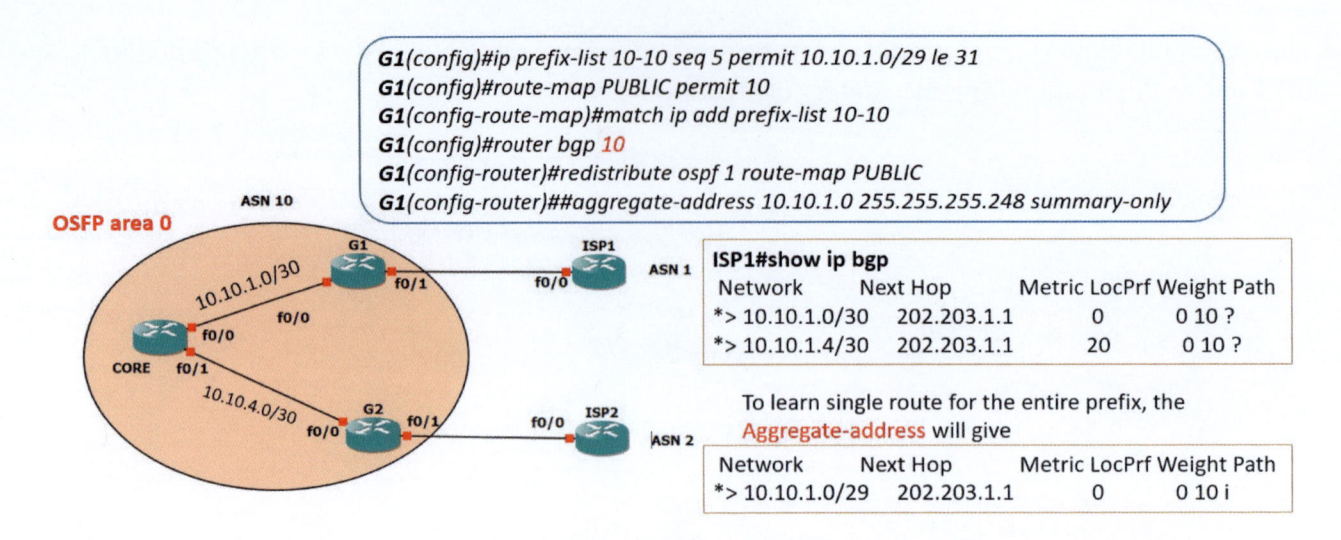

```
G1(config)#ip prefix-list 10-10 seq 5 permit 10.10.1.0/29 le 31
G1(config)#route-map PUBLIC permit 10
G1(config-route-map)#match ip add prefix-list 10-10
G1(config)#router bgp 10
G1(config-router)#redistribute ospf 1 route-map PUBLIC
G1(config-router)##aggregate-address 10.10.1.0 255.255.255.248 summary-only
```

ISP1#show ip bgp

Network	Next Hop	Metric	LocPrf	Weight	Path
*> 10.10.1.0/30	202.203.1.1	0		0	10 ?
*> 10.10.1.4/30	202.203.1.1	20		0	10 ?

To learn single route for the entire prefix, the **Aggregate-address** will give

Network	Next Hop	Metric	LocPrf	Weight	Path
*> 10.10.1.0/29	202.203.1.1	0		0	10 i

Figure 9.49 advertising inbound route to the internet

Satellite links

Satellite communication has revolutionized the telecommunication industry since 1960s. They made possible (and still today) connection to remote regions where mountainous and other terrain obstacles pose challenges to microwave and fiber deployment. However there are disadvantages of satellite link in mobile communication

- The satellite link is very long with the latency associated very high. UMTS and LTE services demand small latency to support VOIP and video streaming. In recent time MEO (medium earth orbit) satellites claim to reduce latency of GEO (geo stationary orbit satellites)
- Satellite links are expensive in terms of spectrum and hence per MB cost
- As operator bandwidth requirement increases, the earth station specification needed increases which adds big margin to the project cost

On the other hand satellite links are easy to install and troubleshoot. Operators who don't have fiber and only relay on microwave for their transmission can use satellite terminals as redundancy to sensitive links. Once VSAT terminals are set up, bandwidth on demand can be used to reduce operational expenditure.

When working on satellite systems and communicating with service providers various terms in the uplink and downlink need to be known as shown in figure 5.50 below

Link design and installation

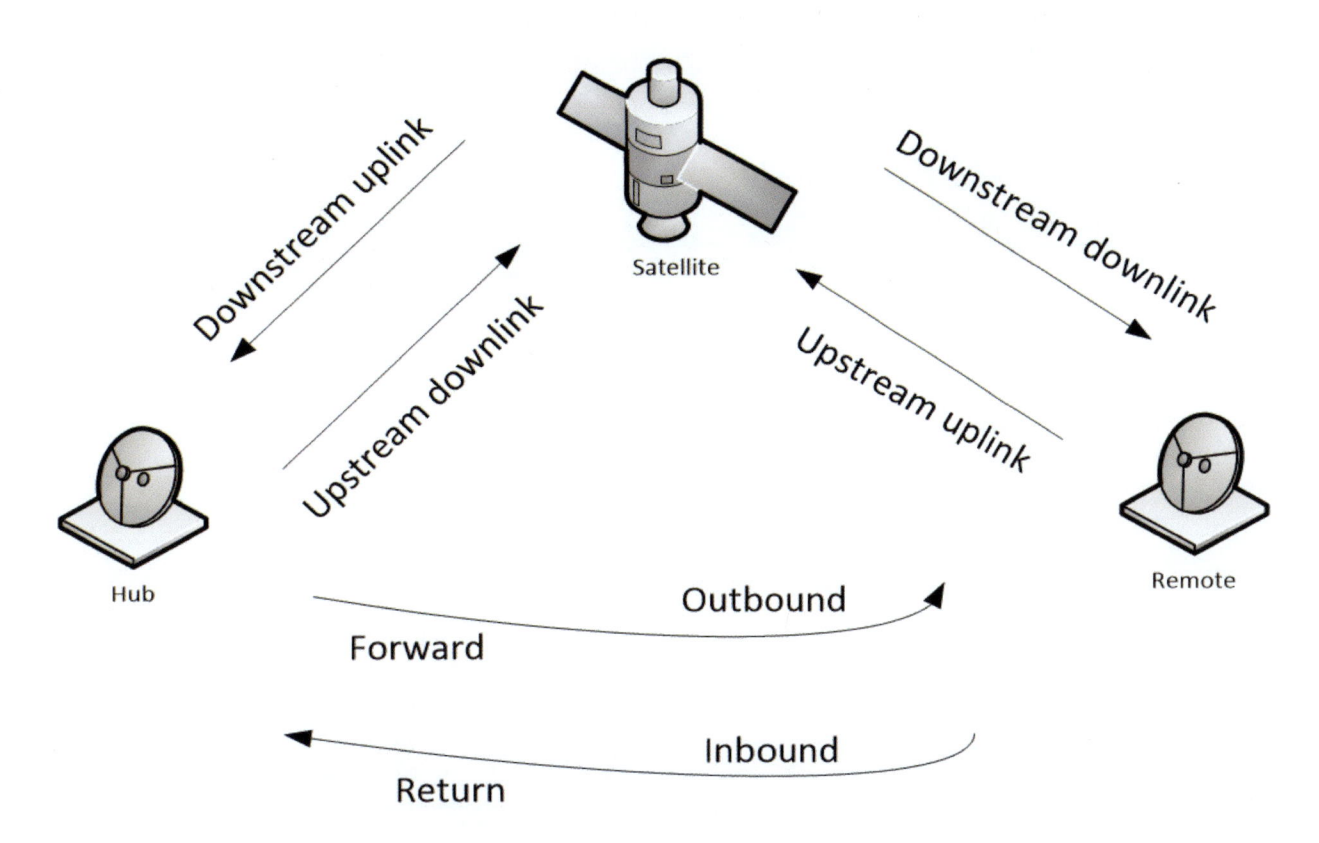

Figure 5.50 satellite signal direction names

In telecommunication satellites operate in different bands such as C, Ku, and Ka. These are summarized in table 5.41 below

Table 5.41 satellite frequency bands

Band	Uplink range (UL)	Downlink range (DL)
C	5.925 – 6.425 GHz	3.7 – 4.2 GHz
Ku	14 – 14.5 GHz	10.7 – 12.25 GHz
Ka	27.5 – 31 GHz	17.7 – 21.2 GHz

Remind in mind that those frequency are RF frequencies transmitted from the BUC or arriving at LNB from satellite. The IF frequency for the baseband modem can be calculated using the local oscillator (LO) frequency of the LNB and BUC as follows

In the UL, the up-converter in the BUC adds the IF frequency from the baseband modem to the LO of the BUC as

$$RF\ frequecy\ to\ the\ satellite\ (UL) = BUC\ LO + IF\ frequency\ from\ modem$$

In the DL, the down-converter in the LNB subtracts the received RF frequency from the satellite from LO of the LNB as

IF frequency to the modem = LNB LO − RF frequency from satellte (DL)

As an example suppose the satellite operators allocates you uplink carrier on 6332 MHz and you purchased 100W BUC of LO 4900 MHz, then the L-band frequency to configure on the modem is

L − band frequency = RF frequency − LO = 6332 − 4900 = 1432 MHz

The IF frequency has frequency range of 70 – 140 MHz in the IF-band and 950 – 2150 MHz in the L-band. If L-band is to be used, cable length and attenuation should be reduced

The IF or L-band is configured on the baseband modem and the RF frequency is configured on BUC in the UL and on LNB in the DL

A VSAT or earth station link contains the following typical component shown in figure 5.51 below

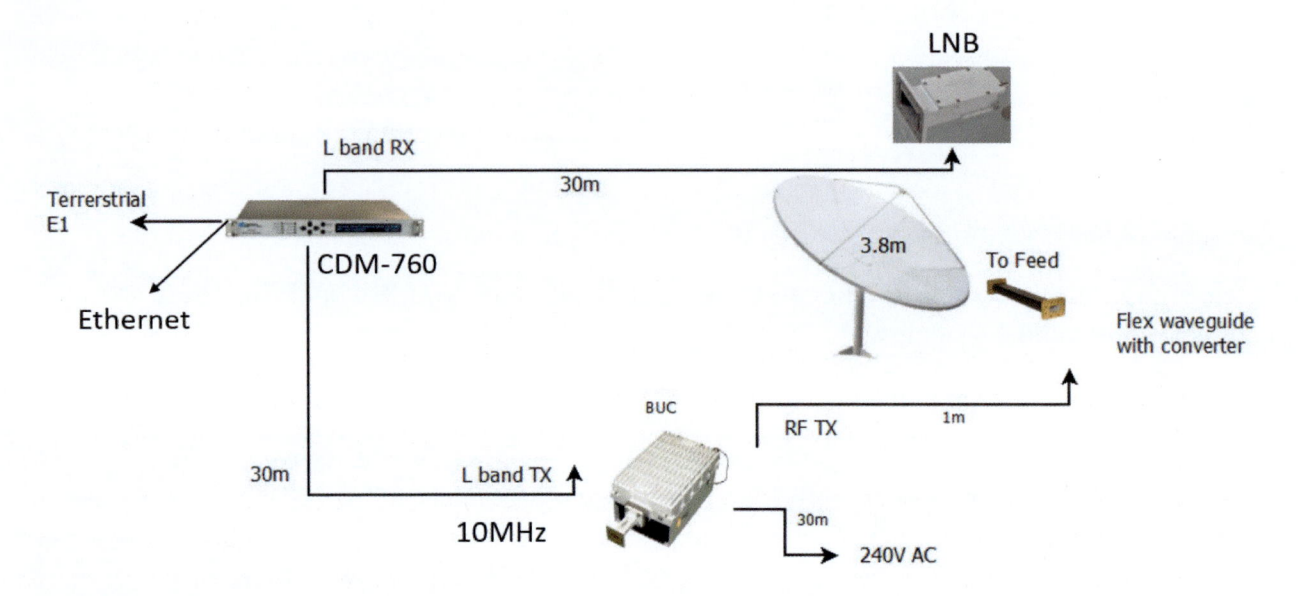

Figure 5.51 VSAT link components

The BUC is transmit equipment in the direction to the satellite. The LNB is the receiver equipment in the direction from the satellite. The modem is the baseband unit that processes the digital bits. The IF cable length puts loss on the link if it is long. These days E1 lines are replaced by Ethernet. For spectrum bandwidth cost to reduce, the size of the dish should be large (>4.5m). Smaller dishes in high wind regions may move out of alignment especially on tall roof.

Like any other wireless communication, link budget should be computed for the satellite VSAT link to determine

- What transmit power is required to produce good reception at the satellite antenna

- What modulation and coding are required to achieve target capacity. This in turn will determine modem capacity to purchase

Satellite links typically use QPSK, BPSK, 8PSK and QAM depending on the required bandwidth. The bandwidth could be dedicated or shared with other customers of the satellite when the bandwidth is not used all the time.

But when purchasing shared bandwidth, things to bear in mind include committed information rate (CIR) and bursting capacity. When the number of customers sharing the bandwidth are many, contention rate will also need to be considered.

Thus the mobile network operator will need consider the following for satellite mobile backhauling

- To reduce the bandwidth cost the possibility of using inclined orbit satellite while weighing pros and cons
- Which band to use. Ku and Ka band are susceptible to rain fading and terrestrial microwave radio links but allow smaller earth station terminals
- Whether to buy shared or dedicated bandwidth

Simplified RF parameters involved in typical VSAT link budget is shown in figure 5.52. We assume bent-pipe (transparent) GEO transponder which just amplifies the signal (no on-board baseband processing)

Pathloss = L
$$L = \left(\frac{4\pi R}{\lambda}\right)^2$$

EIRP = (TX power) x (TX antenna gain)

Link performance at receiver input = C/ No

Receiver equipment performance (figure or merit) = G/T

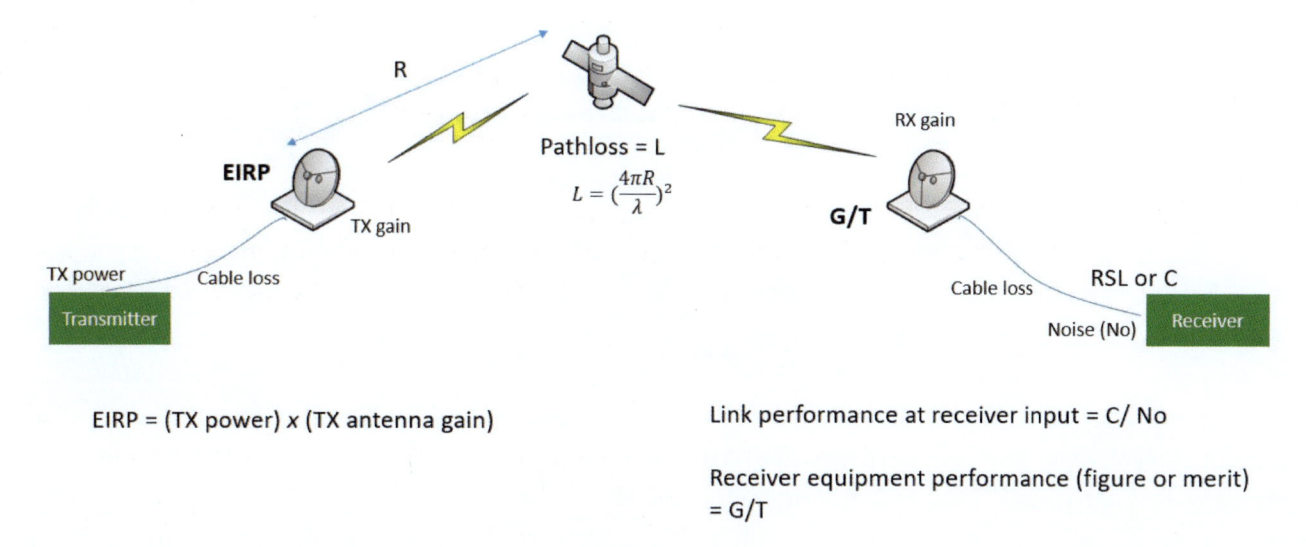

Figure 5.52 simplified VSAT link budget parameters

Example – uplink budget

Company A wants to establish a C-band link. They want to design the link with following parameters
Frequency = 6GHz, antenna size = 4.5m, transmit power = 20dB, geostationary orbit of 40,000km,
Antenna gain = 50dBi

EIRP = 20 + 50 = 70dBW

Pathloss = $L = (\frac{4\pi R}{\lambda})^2 = (\frac{4\pi R f}{c})^2 = (\frac{4\pi R f}{c})^2 = (\frac{4\pi (40,000,000) 6,000,000,000}{300,000,000})^2 = 200\text{dB}$

RSL = EIRP – all losses + receive antenna gain
RSL = 70 – 200 + 50 = -70dBm

Example – downlink budget

Company A wants to establish a C-band link. They want to design the link with following parameters
Frequency = 4GHz, antenna size = 4.5m, transmit power = 20dB, geostationary orbit of 40,000km,
Antenna gain = 50dBi

EIRP = 20 + 50 = 70dBW

Pathloss = $L = (\frac{4\pi R}{\lambda})^2 = (\frac{4\pi R f}{c})^2 = (\frac{4\pi R f}{c})^2 = (\frac{4\pi (40,000,000) 4,000,000,000}{300,000,000})^2 = 190\text{dB}$

RSL = EIRP – all losses + receive antenna gain
RSL = 70 – 190 + 50 = -60dBm

We can also define link performance between the Earth station and the Satellite input receive antenna as follows. The link performance at the receiver input is defined as ratio between the receiver RSL or C and noise spectral density (No) ➔ C / No

Link design and installation

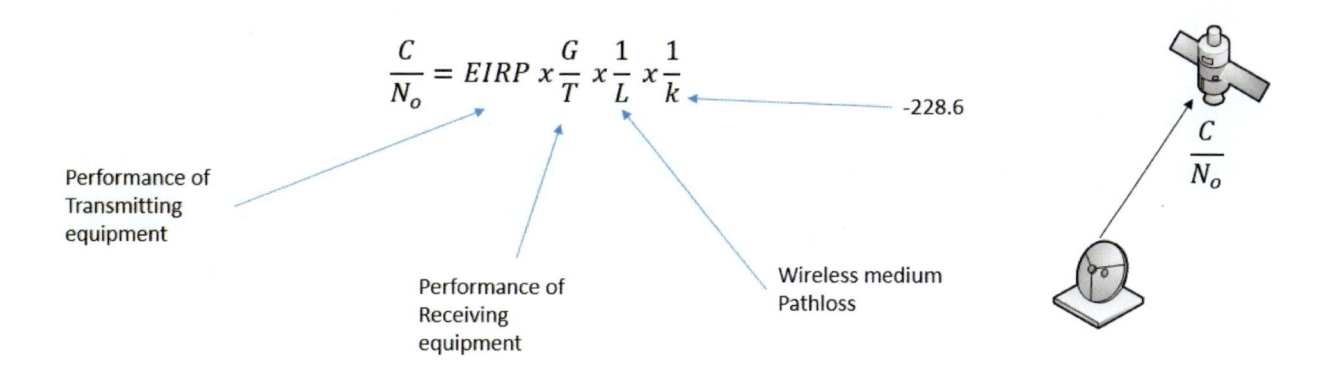

$$\frac{C}{N_o} = EIRP \times \frac{G}{T} \times \frac{1}{L} \times \frac{1}{k}$$

- Performance of Transmitting equipment → EIRP
- Performance of Receiving equipment → G/T
- Wireless medium Pathloss → 1/L
- −228.6 → 1/k

We need also define the overall link performance (Earth station to Earth station) assuming bent-pipe transponder that just amplify and downcovert the received signal

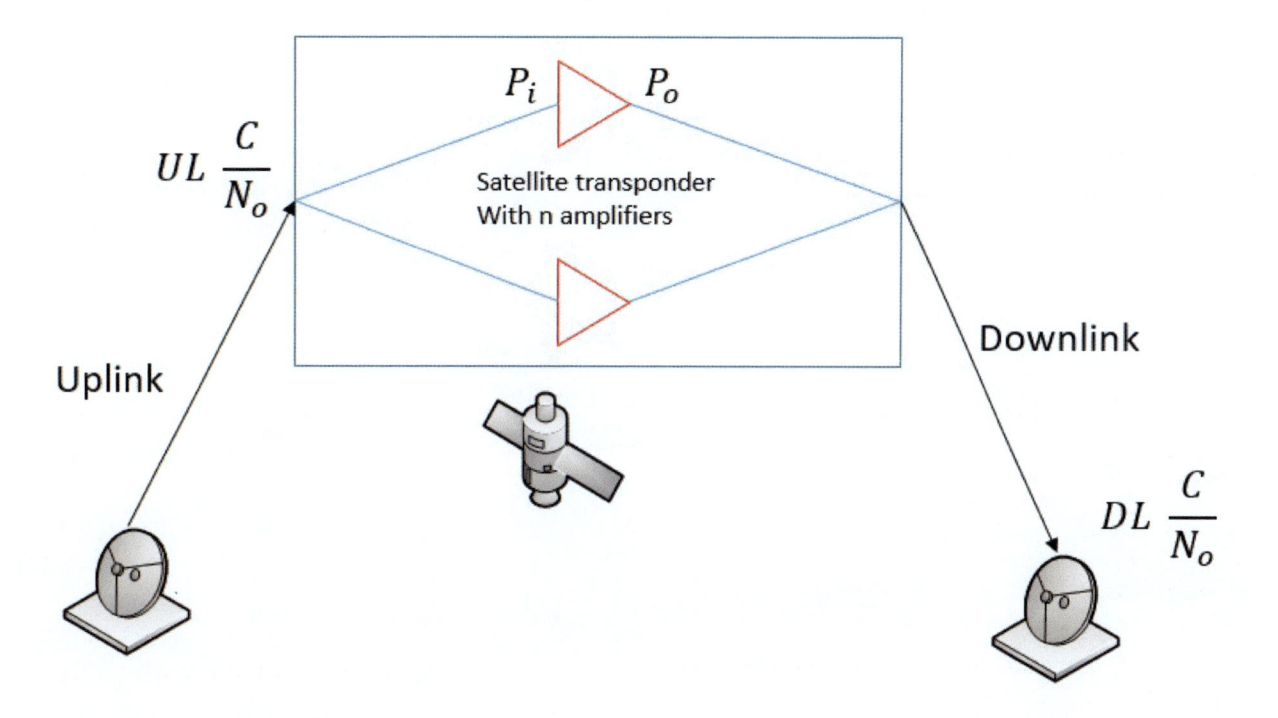

Figure 5.53 Overall link performance

$$UL \frac{C}{N_o} = input\ backoff + UL \frac{C}{N_o} at\ saturation$$

$$DL \frac{C}{N_o} = output\ backoff + DL \frac{C}{N_o} at\ saturation$$

Input backoff = operating input power to the Satellite channel – saturated input power to the satellite channel = Pi - Pisat

Output backoff = operating output power of the Satellite channel – saturated output power of the satellite channel = Po - Posat

The above formulas assume not interference on the overall link

An example link budget of 50Mbps internet from Mogadishu hub to Bosaso station is shown in table 5.15

Table 5.15 Example of satellite link budget (should not be used in real network)

Link Budget
Produced using Satmaster Pro
Saturday 14 March 2020

Service Name	Internet backhaul to bosaso		
Coverage	100		
Uplink earth station	MOGADISHU252		
Downlink earth station	BOSASO252		
Satellite name	IS 902		
Modcod	Manual		

Link Input Parameters	**Up**	**Down**	**Units**
Site latitude	2.0469N	11.2755N	degrees
Site longitude	45.3182E	49.1879E	degrees
Site altitude	0.000	0.491	km
Frequency	6	4	GHz
Polarization	Circular	Circular	
Rain model	ITU-R	ITU-R	
Rain zone or R0.01% (mm/h)	50.399	6.446	
Availability (average year)	99.9000	99.9000	%
Antenna aperture	4.5	4.5	metres
Antenna efficiency or gain (+ or - prefix)	70	70	% or dBi
Coupling loss	1	1	dB
Antenna mispoint loss	1	2	dB
Other path losses (site diversity gain -ve)	2	1	dB
LNB noise figure or temp (+ prefix)		0.6	dB or K
Antenna noise		32.25	K
C/ACI	12		dB
C/ASI	-16.53		dB
C/CCI	15		dB
HPA C/IM	50		dB
C/ACI		12	dB
C/ASI		-90.93	dB
C/CCI		15	dB
Uplink station HPA output back-off	1		dB
Uplink power control available	0		dB
Number of carriers / HPA	1		
Required HPA power	40		W

Satellite Input Parameters	Value	Units
Satellite longitude	62.00E	degrees
Transponder type	LTWTA	
G/T Reference	1.8	dB/K
SFD Reference	-90	dBW/m2
Receive G/T	-1.5	dB/K
Attenuator pad (gain step)	1	dB
Effective SFD	**-85.70**	**dBW/m2**
Satellite ALC	1	dB
EIRP (saturation)	4	dBW
EIRP (beam peak)	25	dBW
Transponder bandwidth	500	MHz
Input back off total	10	dB
Output back off total	AUTO	dB
C/IM	30.00	dB
Carriers per transponder	AUTO	

Carrier/Link Input Parameters	Value	Units
Modulation	4-PSK	
Required Eb/No	10	dB
Information rate	50	Mbps
Information rate overhead	20	%
FEC code rate	0.75	
Spreading gain	0	dB
(1 + Roll off factor)	1.45	
Carrier spacing factor	2	
Bandwidth allocation step size	5	MHz
Implementation loss	0	dB
System margin	1	dB

Calculations at Saturation	Value	Units
Gain 1m^2	37.02	dB/m2
Uplink C/No	104.38	dB-Hz
Downlink C/No	55.73	dB-Hz
Total C/No	55.73	dB-Hz
Uplink EIRP for saturation	79.49	dBW

General Calculations	Up	Down		Units
Elevation	70.27	70.04		degrees
True azimuth	96.80	130.69		degrees
Compass bearing	97.10	129.85		degrees
Path distance to satellite	36103.45	36111.05		km
XPD during rain	122.57	229.81		dB
Propagation time delay	0.120428	0.120453		seconds
Antenna efficiency	70.00	70.00		%
Antenna gain	47.48	43.96		dBi
Availability (average year)	99.9000	99.9000		%
Link downtime (average year)	8.766	8.766		hours
Availability (worst month)	99.6154	99.6154		%
Link downtime (worst month)	2.809	2.809		hours

Uplink Calculation	Clear	Rain Up	Rain Dn	Units
Transmit EIRP	61.51	61.51	61.51	dBW
Uplink power control used	0.00	0.00	0.00	dB
Transponder input back-off (total)	10.00	10.00	10.00	dB
Input back-off per carrier	17.99	18.48	17.99	dB
Antenna mispoint	1.00	1.00	1.00	dB
Free space loss	199.16	199.16	199.16	dB
Atmospheric absorption	0.05	0.05	0.05	dB
Tropospheric scintillation	0.00	0.18	0.00	dB
Cloud attenuation	0.00	0.06	0.00	dB
Rain attenuation	0.00	0.39	0.00	dB
Total attenuation (gas-rain-cloud-scintillation)	0.05	0.54	0.05	dB
Other path losses	2.00	2.00	2.00	dB
C/No (thermal)	86.40	85.91	86.40	dB-Hz
C/N (thermal)	10.37	9.88	10.37	dB
C/ACI	12.00	11.51	12.00	dB
C/ASI	-34.52	-35.01	-34.52	dB
C/CCI	15.00	15.00	15.00	dB
C/IM	50.00	50.00	50.00	dB
C/(N+I) [Es/(No+Io)]	-34.52	-35.01	-34.52	dB
Eb/(No+Io)	-36.28	-36.77	-36.28	dB

Downlink Calculation	Clear	Rain Up	Rain Dn	Units
Satellite EIRP total	4.00	4.00	4.00	dBW
Transponder output back-off (total)	7.02	7.02	7.02	dB
Output back-off per carrier	15.01	15.01	15.01	dB
Satellite EIRP per carrier	-11.01	-11.01	-11.01	dBW
Antenna mispoint	2.00	2.00	2.00	dB
Free space loss	195.64	195.64	195.64	dB
Atmospheric absorption	0.04	0.04	0.04	dB
Tropospheric scintillation	0.00	0.00	0.11	dB
Cloud attenuation	0.00	0.00	0.01	dB
Rain attenuation	0.00	0.00	0.00	dB
Total attenuation (gas-rain-cloud-scintillation)	0.04	0.04	0.16	dB
Other path losses	1.00	1.00	1.00	dB
Noise increase due to precipitation	0.00	0.00	0.03	dB
Downlink degradation (DND)	0.00	0.00	0.14	dB
Total system noise	130.37	130.37	131.13	K
Figure of merit (G/T)	21.81	21.81	21.79	dB/K
C/No (thermal)	40.72	40.72	40.58	dB-Hz
C/N (thermal)	-35.30	-35.30	-35.44	dB
C/ACI	12.00	11.51	12.00	dB
C/ASI	-105.94	-105.94	-105.94	dB
C/CCI	15.00	15.00	15.00	dB
C/IM	30.00	30.00	30.00	dB
C/(N+I) [Es/(No+Io)]	-105.94	-105.94	-105.94	dB
Eb/(No+Io)	-107.70	-107.70	-107.70	dB

Totals per Carrier (End-to-End)	Clear	Rain Up	Rain Dn	Units
C/No (thermal)	40.72	40.72	40.58	dB-Hz
C/N (thermal)	-35.30	-35.30	-35.44	dB
C/ACI	8.99	8.50	8.99	dB
C/ASI	-105.94	-105.94	-105.94	dB
C/CCI	11.99	11.99	11.99	dB
C/IM	29.95	29.95	29.95	dB
C/I (total)	-105.94	-105.94	-105.94	dB
C/(No+Io)	-29.92	-29.92	-29.92	dB-Hz
C/(N+I) [Es/(No+Io)]	-105.94	-105.94	-105.94	dB
Eb/(No+Io)	-107.70	-107.70	-107.70	dB
Implementation loss	0.00	0.00	0.00	dB
System margin	1.00	1.00	1.00	dB
Net Eb/(No+Io)	-108.70	-108.70	-108.70	dB
Required Eb/(No+Io)	10.00	10.00	10.00	dB
Excess margin	**-118.70**	**-118.70**	**-118.70**	dB

EIRP Density Calculations	Clear	Rain Up	Rain Dn	Units
Flange transmit (up)	-62.00	-62.00	-62.00	dBW/Hz
Satellite (down)	-87.03	-87.03	-87.03	dBW/Hz
Satellite beam peak (down)	-66.03	-66.03	-66.03	dBW/Hz
Flange receive (down)	-241.74	-241.74	-241.89	dBW/Hz
Antenna off axis transmit toward 61E	-28.61			dBW/Hz

Earth Station Power Requirements	Value	Units
EIRP per carrier	61.51	dBW
Available uplink power control	0.00	dB
Total EIRP required	61.51	dBW
Antenna gain	47.48	dBi
Antenna feed flange power per carrier	14.02	dBW
HPA output back off	1.00	dB
Waveguide loss	1	dB
Number of HPA carriers	1	
Total HPA power required	16.0206	dBW
Required HPA power	40.0000	W

Space Segment Utilization	Value	Units
Overall availability	99.8001	%
Information rate	50.0000	Mbps
Information rate (inc overhead)	60.0000	Mbps
Transmit rate	80.0000	Mbps
Symbol rate	40.0000	Mbaud
Noise Bandwidth	76.02	dB-Hz
Occupied bandwidth	58.0000	MHz
Minimum allocated bandwidth required	80.0000	MHz
Allocated transponder bandwidth	80.0000	MHz
Overall Link efficiency	0.750	bps/Hz
Percentage transponder bandwidth used	**16.00**	**%**
Used transponder power	-11.01	dBW
Percentage transponder power used	**15.90**	**%**
Max carriers / transponder	6.25	
Limited by:	Bandwidth	
Power equivalent bandwidth usage	79.5128	MHz

From this link budget the satellite operator connecting the two sites reported the following link specification (table 5.16) and earth station equipment for the sites (table 15.17)

Table 5.16 link description

Name	Mogadishu to Bosaso 50Mb internet
Teleport location	2.045N, 45.318E (Mogadishu)
Remote site location	11.28N, 49.18E (BOSASO)
Satellite	Intelsat 902
Transponder	10/10
Polarization	L for transmit/R for receive
Capacity	50MB

Table 5.17 earth station the customer should order

Antenna model	Prodelin 7.2m
BUC model	Agilis 100W
BUC LO	4900
LNB model	Norsat 3220F
LNB LO	5150MHz
Modem	Newtech MDM 6000

Now the equipment arrives and the VSAT engineer first works with civil engineer to prepare the concrete slab to support the antenna.

Installation should be done carefully and proper coordination with satellite support center should be maintained by the VSAT engineer. During installation the beacon signal of the satellite is searched with spectrum analyzer. Alternatively if spectrum analyzer is not available a television receiver could be used and quality of free to air channels are observed.

The following table illustrates a typical list the VSAT engineer is required to check before finalizing the project

Table 5.18 Satellite link typical checklist

Observe beacons on analyzer?	Confirmed R polarization receives. Detected satellite
Satellite detected carrier transmission?	YES. Pointing correct
CW of remote carrier detected on analyzer?	YES. Nominal with BER of zero
Feed horn Polarization correct?	YES. Isolation test with operator confirmed high CPI reported
Carrier TX power nominal?	YES. Pure carrier transmission to Satellite confirmed and stabilized
Link stable	YES. 24 hours continuous monitoring confirmed and performance report generated

BER	Nominal
RSL	Nominal
Iperf bandwidth test	Ok

Like other wireless systems, satellite radio links can fail due to various reason. If it happens a VSAT engineer gets alarm of link outage, some of the things he would typically due include

- Check transmit and receive carriers status with Satellite support team
- Check RX chain. Cables, connectors, LEDs, LNB, loose connections
- Make sure configurations match on both ends
- Adjust power levels with satellite support team
- Access BUC via browser and check SSPA, frequency entry, and other indicators
- Power cycle LNB and check the 18V from modem with voltage meter
- Reset modem or BUC in case the system fails to respond to connectivity
- Does teleport use same frequency range? Check IF or L band
- Ask teleport if they can receive your signal in continuous wave (CW)
- If you lost RX signal, verify if teleport can receive its own signal or loopback itself (link between teleport and satellite is ok)
- Confirm if they get lock at RF level. Check if baseband is not getting synchronized
- Check symbol rate of your TX and teleport RX. Symbol rate is result of several factors including FEC, modulation, data rate. If one of this is not matched you will see different symbol rate
- Check spectrum inversion and try to toggle it on off
- Disable and then enable interfaces
- Check moisture in the feedhorn and clean antenna system periodically. If it is tracking antenna system, proper lubrication should be periodically applied to the moving parts
- Replace LNB or BUC, or modem as last resort when all other step don't produce any change

We will take one case in which we will illustrate a VSAT junior engineer and how he handled with link outage as explained in the box below

Link design and installation

The tense day of the junior VSAT engineer

One day a VSAT engineer receives a mail alert that the internet gateway is down. The operator gateway is 7.2 standard focus dish connected to remote teleport in Germany. It is 12AM and traffic is at peak.

The VSAT engineer arrives on the site and the first thing he does is that he re-routes the traffic to another **standby link**. In that case the operator always has redundancy. The VSAT engineer arrives the site with link design, layout files and tool box. In this case **he knows what it is required to accomplish his job**

He calls the ISP team and they tell him that they can reach (ping) the modem public IP interface facing the NAT (network address translation) router. This **isolates** the problem starting from modem to the antenna system.

He goes to the VSAT indoor equipment rack and first thing he inspects is system LEDs. He finds out that both TX and RX LEDs on the modem are green (normal). He also tries to tighten all connectors. In the meantime he is also **communicating** with the remote teleport support team

The **remote support team** gives him further information to help him understand more. They tell him that his carrier is not received on the satellite and they cannot ping his modem interface facing them

The VSAT engineer **reasons** that if modem LEDs indicate TX and RX green, then maybe the problem is not LNB, BUC or IF cables.

Now he accesses the modem configuration screen to check any errors. Since his receive is ok, he opens the transmit configuration section. He first mutes and unmutes the carrier and no effect is seen. The link is still down. The remote teleport modem is still unlocked.

Now with 1 hour precious time gone and customers complaining, he **escalates** the situation to his senior supervisor. The senior supervisor checks the configuration and finds out that the TX carrier is enabled on the modem, but the external 10MHz clock is not provided to the BUC. With the clock activated the link comes up as reported by the remote teleport.

Lessons learned

- Responding to the problem with clear plan and strategy
- Communication and proper coordination
- Problem isolation to narrow it down to specific area
- First trying resolution, then problem escalation to senior team

Chapter Six

Frequency and capacity planning

Frequency planning is an important step when designing microwave radio links. One of the early steps of the design is the selection of suitable frequency band. It is assumed that a microwave network operator will perform frequency scanning in the pre-planning phase, mark channels that is not used, and license them. Alternatively, Communication commission of the country will have already done spectrum monitoring and measurement and published all used and available bands.

After route design is established and frequency bands selected, a suitable channel arrangement will be selected based on capacity requirements.

Route design → Link design → Frequency planning

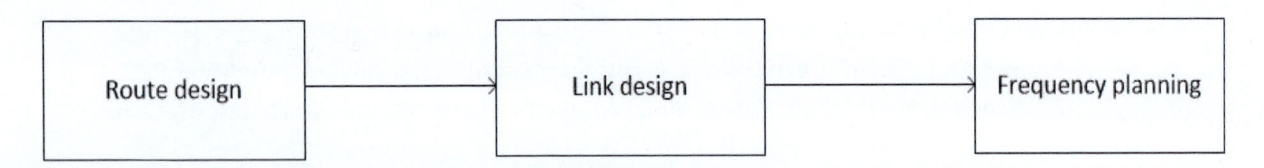

Figure 6.1 Microwave network design steps

One thing to remember when evaluating capacity requirements is, the operator should specify what services it will provide. Each service requires bandwidth and hence capacity

Table 6.1 examples of telecommunication services

Services	Plan
Voice calls	
Voice over IP	
Internet access	All services demand bandwidth from the network
Video conferencing	
Online gaming	
Management data	

Once service bandwidth requirements is established, a suitable link bandwidth should be selected. Microwave link capacity is decided by three RF parameters

- Radio channel bandwidth
- Modulation order

Frequency and capacity planning

- Radio frame transmission efficiency.

Microwave LOS frequency bands and channel arrangement is already developed by ITU-R F series documentations. It is published on ITU-R website. A case study will be illustrated in appendix B

First step is allocation of specific frequency band to Microwave services called **frequency allocation** in ITU, then channel plans is designed under the given band in a process called **frequency allotment.** Finally frequencies in the channel plans are assigned to radio stations in a process called **frequency assignment**. This process is illustrated in the figure 6.2

For countries that follow European ETSI recommendations normally use frequency multiples of 3.5 MHz while for American ANSI typically employ multiple of 2.5 MHz in channel plans

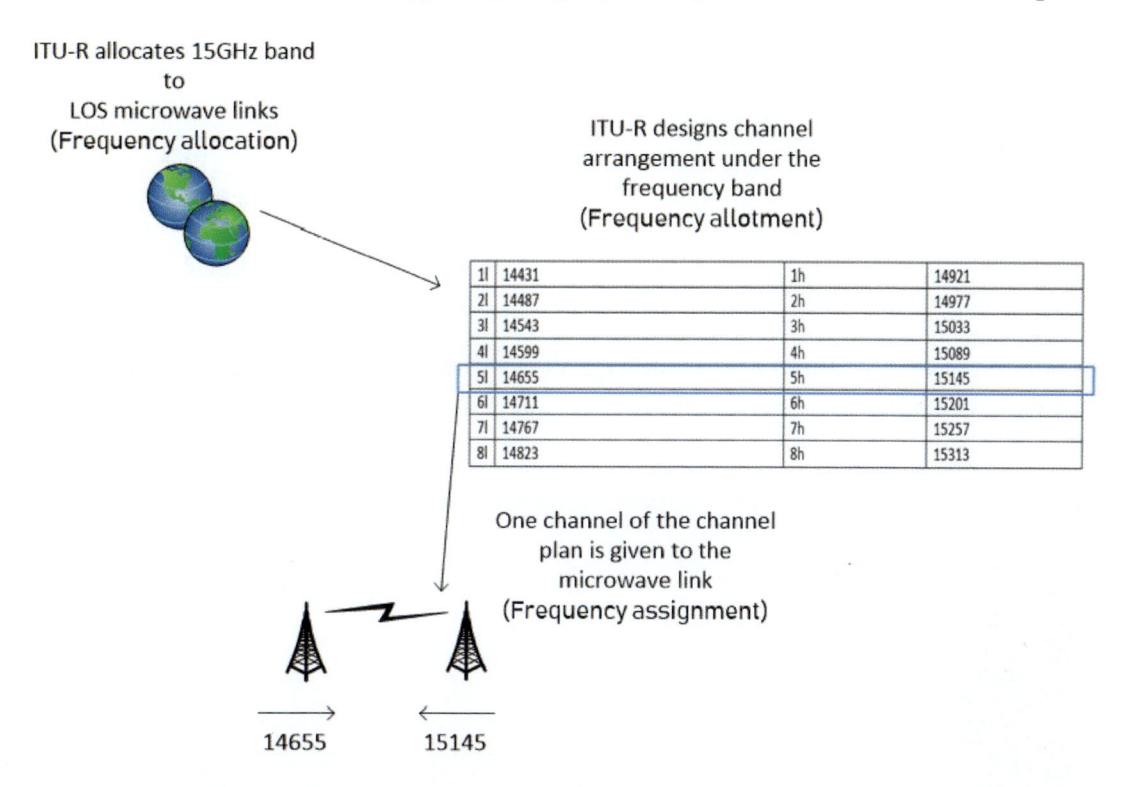

Figure 6.2 Difference between frequency allocation, allotment, assignment

The example in the figure shows allocation of 15GHz band to microwave services.

ITU-R allocated frequency bands to LOS microwave links and is summarized in the following table

Table 6.2 ITU-R Microwave LOS frequency allocation

Band	Range	Application
L6	5.9 – 6.4	Long distance backbone routes
U6	6.4 – 7.1	

7	7.1 – 7.7	
8	7.7 – 8.5	
11	10.7 – 11.7	
13	12.7 – 13.2	Short distance mobile backhaul links
15	14.5 – 15.3	
18	17.7 – 19.7	
23	21.1 – 23.6	
26	24.5 – 26.5	
38	37 – 39.5	
42		
60 – 90	70GHz (71-76) 80GHz (81-86)	

To better understand the channel arrangement process, a given frequency band (say 15GHz) is divided into two equal parts as shown below

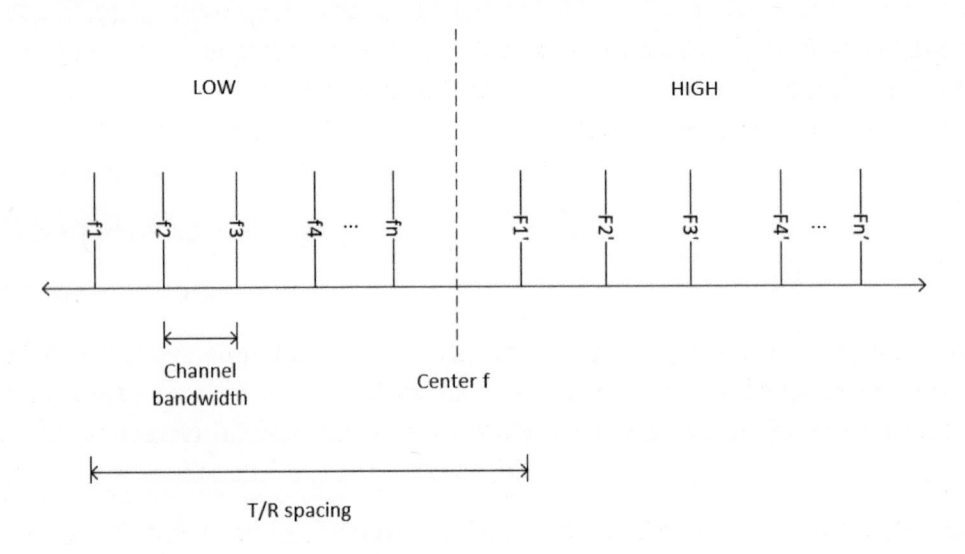

Figure 6.3 ITU-R radio channel plan arrangement

The center frequency in the middle of the plan divides the plan into low sub-band and high sub-band. The parameters illustrated in the channel plan are explained as follows

- Center frequency is the frequency that divides the channel plan into two equal parts.

Frequency and capacity planning

- Channel bandwidth is the difference between any two adjacent frequencies either in the LOW sub-band or the HIGH sub-band.
- T/R spacing is the duplex spacing. It is the difference between channel in the LOW sub-band and its corresponding frequency in the HIGH sub-band.

Take 15GHz band as an example. The range of the 15GHz band is 14.5 to 15.35 GHz

Center frequency of the band is (14.5 + 15.35)/2 which is equal to 14.925

If we need high capacity links, we can choose 56 MHz bandwidth channel plan. Then following ITU-R guidelines we will get the following plan

Table 6.3 Example 15GHz band 56MHz channel plan

1l	14431	1h	14921
2l	14487	2h	14977
3l	14543	3h	15033
4l	14599	4h	15089
5l	14655	5h	15145
6l	14711	6h	15201
7l	14767	7h	15257
8l	14823	8h	15313

LOW sub-band of the frequency is given in columns designated as (1l, 2l, and 3l up to 8l). First frequency in the LOW part is 1l, second frequency is 2l, and third frequency is 3l and so on.

Channel bandwidth is the difference between any two adjacent frequencies and should be constant across the whole plan.

Channel bandwidth is the difference (2l – 1l) or (3l – 2l) or (4l – 3l) and so on and it is 56MHz in our example. Similar argument applies to HIGH sub-band.

T/R spacing is the difference (1h – 1l) or (2h – 2l) or (3h – 3l) and so on and it is 490MHz in our example.

Finally when assigning frequencies, the LOW half is used for the transmit station and the HIGH half for the receive station. This means if we have a microwave link connecting two sites, one site is given transmit (LOW) and the second site receive (HIGH). Single site cannot be both LOW and HIGH as it will result HIGH/LOW interference.

When using radio channel plans, under a given frequency band various arrangement are possible based on bandwidth. This means under 15GHz band, different channel plans is obtained based on

Frequency and capacity planning

whether bandwidth is 7 MHz, 14 MHz or 28 MHz and is in multiples of either 2.5 MHz or 3.5 MHz as shown in the table below

Table 6.4 2.5MHz (ANSI) and 3.5 MHz (ETSI) multiples of channel bandwidths

Multiples of bandwidth x 2	Channel plans
2.5 MHz	2.5, 5, 10, 20, 40, 80, 160MHz
3.5 MHz	3.5, 7, 14, 28, 56, 112MHz

So if we have 15GHz, we will have channel plan for 3.5MHz bandwidth, a different channel plan for 7MHz bandwidth, a different channel plan for 14MHz bandwidth and so on. This means two adjacent 28MHz channels can be aggregated to produce single 56MHz center frequency for high capacity links

As the bandwidth of the plan increases (say from 28MHz to 56MHz), more data can be carried by the microwave link (data rate increases).

The following general guidelines are applicable when selecting frequency band for a particular link distance and application

- High frequency bands (13/15/18/23/26/38/42/56) are used for short distance links in metro areas
- Lower frequency bands (L6/U6/7/8/10/11) are used for long distance backbone routes
- Higher bands yield higher data rates under smaller antennas. Short towers are possible here.
- Lower bands for backbone links will require taller towers to provide enough clearance above terrain and large antennas to increase system gain that will compensate large pathloss involved.
- Since most operators prefer main roads connecting cities as route, interference will be more prevalent when many towers from the operators are placed in close proximity.

The effects of excess clearance due to taller towers in lower bands were covered in chapter four (route design). Interference analysis will be covered later in chapter seven.

When design radio channel arrangement for frequency bands, three approaches are taken

- ACAP (adjacent channel alternate polarization) in which adjacent channels are transmitted on opposite polarization.
- ACCP (adjacent channel co-polarization) in which adjacent channels are transmitted in the same polarization (not used due spectrum inefficiency and hence will not be discussed in this text)
- CCDP (co-channel dual polarization) in which every channel is re-used in both polarizations.

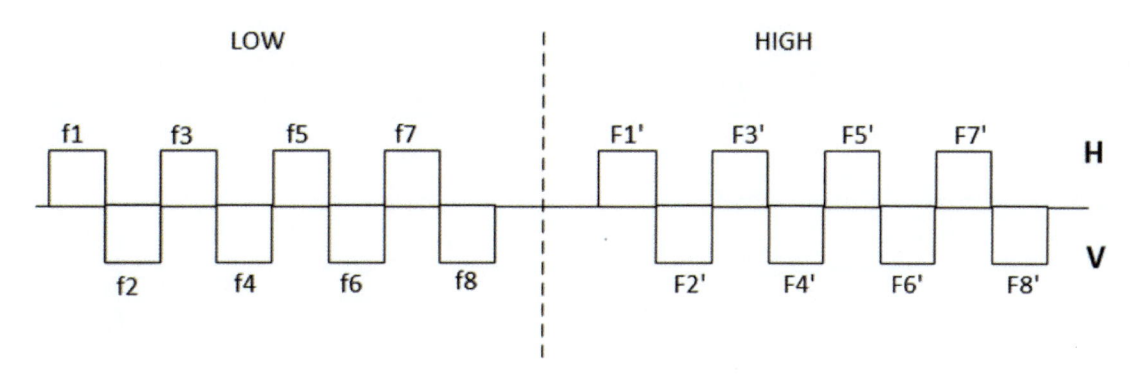

Figure 6.4 ACAP

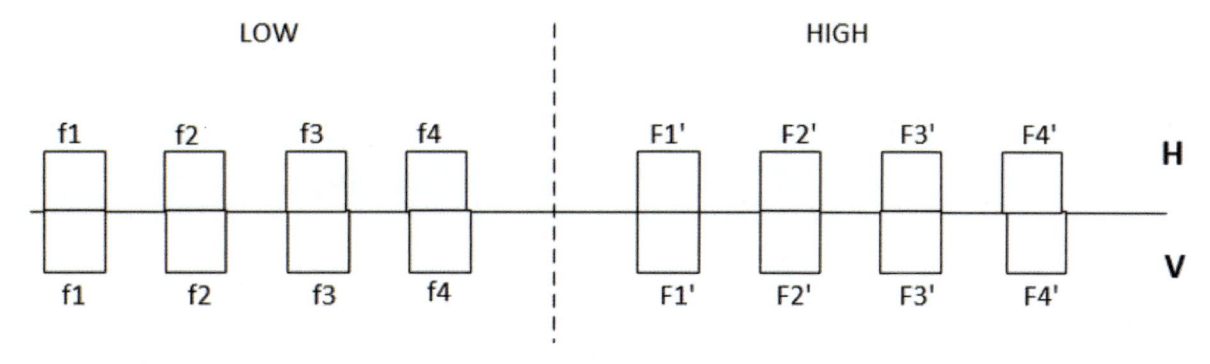

Figure 6.5 CCDP

A good frequency plan will assign fewer frequencies to links as possible while minimizing interference. In backbone links where hop lengths are longer than 40km, it is good practice to use two frequency plan where possible and use alternated polarization in every two hops. If two frequency plan is not sufficient (links interfere each other and not follow zigzag pattern), then four frequency plan could be used. If links are shorter under 30km, then using high performance antennas (with higher F/B ratio) will reduce interference.

Two frequency plan means the use of a pair of frequencies for the entire route.

Table 6.5 two frequency plan

LOW	HIGH
f1	f2

The limitations of two frequency plan is increased interference at the node when it receives same frequency from the stations on either side. If we make hops longer, then the additional attenuation of the carrier will weaken the signal before it can reach adjacent station and cause interference.

Frequency and capacity planning

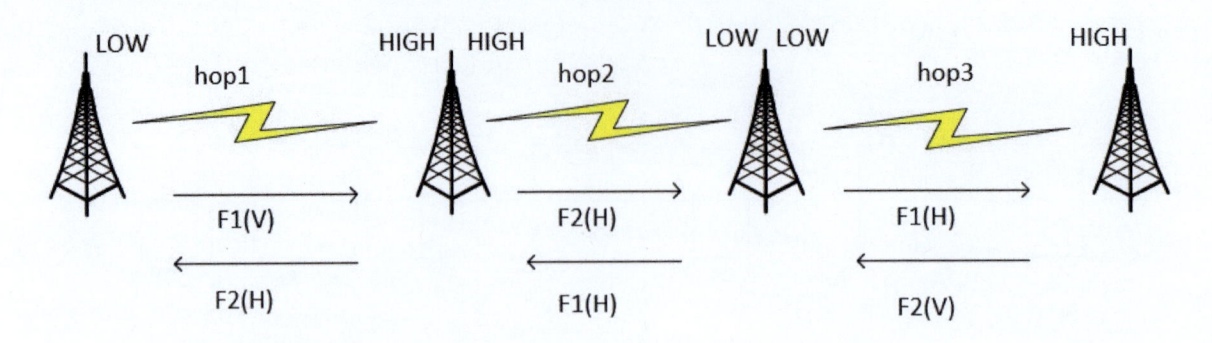

Figure 6.6 Two frequency plan example

The disadvantage of four frequency (four carriers) plan is it requires one more channel license compared with two frequency plan but provides better protection against interference.

Table 6.6 four frequency plan

LOW	HIGH
f1	f2
f3	f4

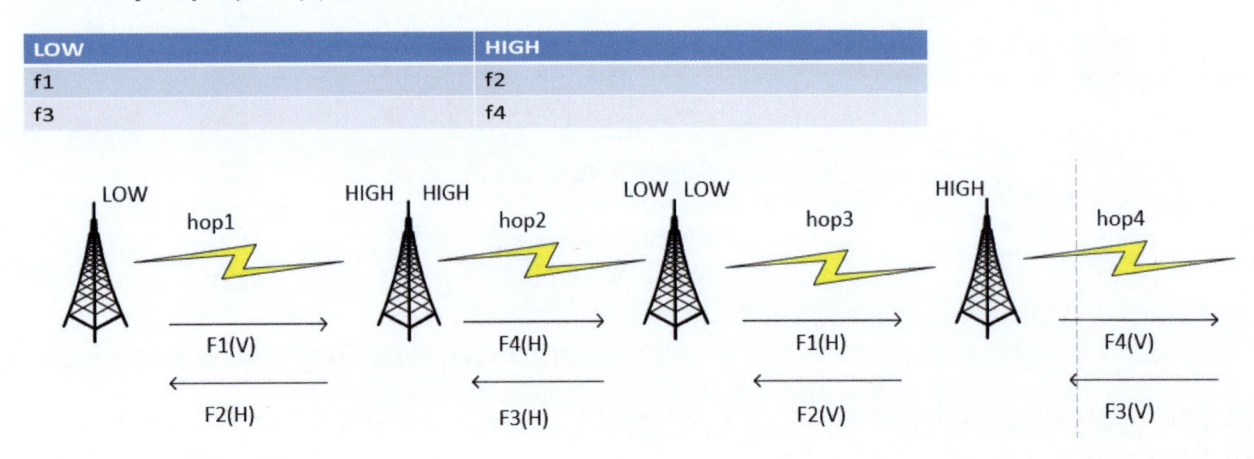

Figure 6.7 Four frequency plan

The example shown above is links in linear topology. For ring topology, if the networks contains even number of stations (2, 4, 6, 8…) then use same frequency band for all the links.

If however there is odd number of stations, then use two different frequency bands. One for the shortest hop and another band for other links. This concept is illustrated in figure 6.8

Frequency and capacity planning

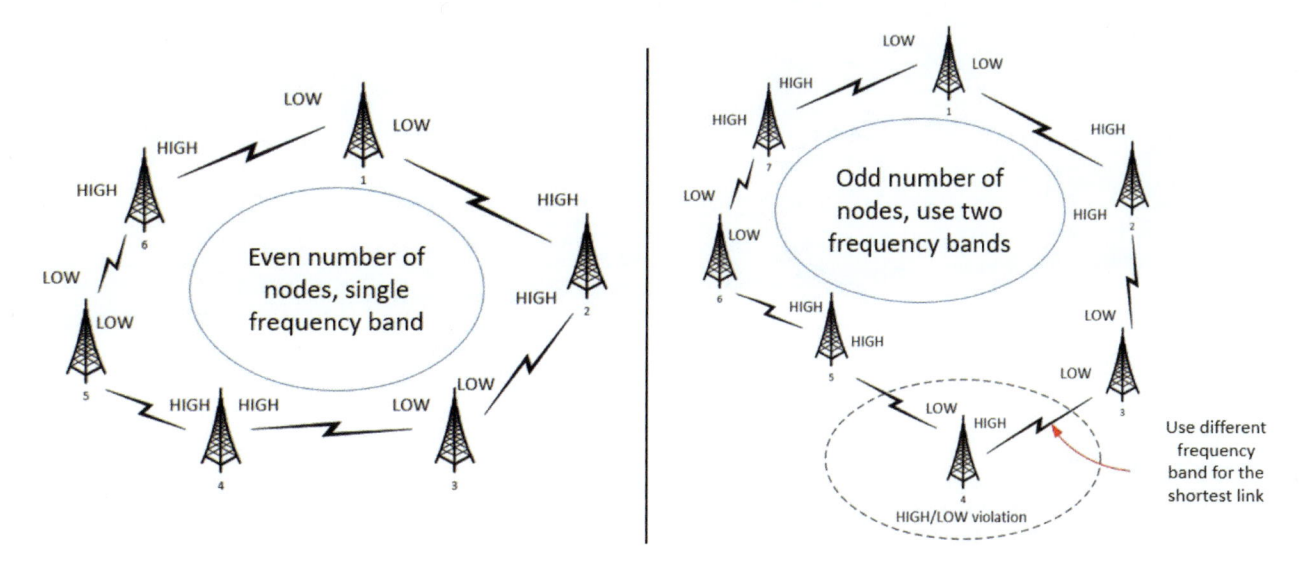

Figure 6.8 Ring frequency planning

So far we have discussed microwave frequency bands that are deployed worldwide links, how frequency allotment and assignment is done, different radio channel plans based on bandwidth, and how frequency are assigned to linear and ring networks.

In this final part of the chapter we will take a case of microwave frequency planning in Pathloss 5.1 tool.

An operator wants to install four links in Mogadishu to form part of larger network route that will connect mobile base stations to core switch located at site A. Due to unavailability of enough power to sites, it was decided the network to take ring topology to provide site redundancy.

Since route and links design are covered in previous chapters, we continue the design with frequency plan in this section and interference analysis in subsequent chapter.

The network diagram is shown below along with frequency assignment to links. Traffic required on each link is shown in table 6.7

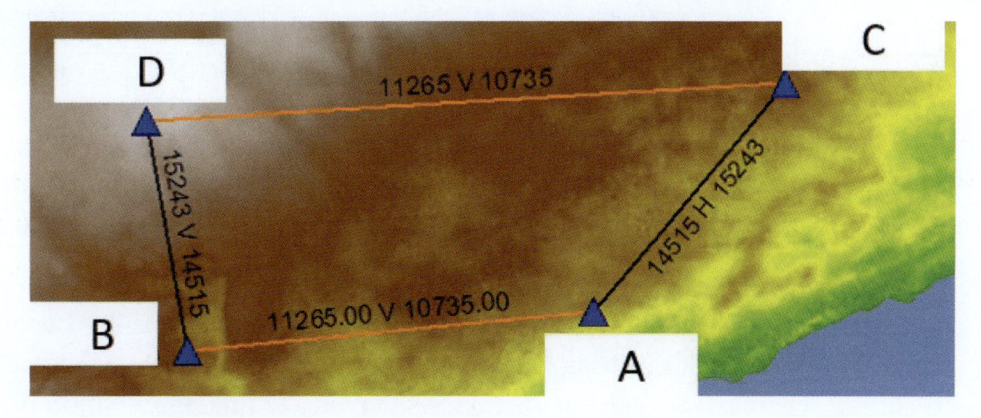

Figure 6.9 ring topology with frequency re-use

	A	B	C	D
A	-	4Gbps	2.5Gbps	-
B	4Gbps	-	-	4Gbps
C	2.5Gbps	-	-	2.5Gbps
D	-	4Gbps	2.5Gbps	-

Table 6.7 traffic matrix

Before selecting a given channel plan, traffic requirements (traffic matrix) of the links must be analyzed. If a lot of data is to be transported by the link, then large channel bandwidth will be required.

The main switch locates at A and traffic aggregation happens there. The other links connect A to other parts of the country.

Traffic matrix and channel bandwidth is shown in the table 6.8

As a general rule, the operator with the support of the vendor has decided to allocate 56 MHz bandwidth to 2.5Gbps links and 112 MHz to 4Gbps links. The allocation of the bandwidth according to traffic matrix is summarized in the following table

Table 6.8 links requirement

Link	Link data rate	Bandwidth	Band	Configuration
A – B	4Gbps	112Mhz	11Ghz	2 + 0 CCDP
A – C	2.5Gbps	56Mhz	15GHz	1 + 0
B – D	4Gbps	112Mhz	15Ghz	2 + 0 CCDP
C – D	2.5Gbps	56Mhz	11Ghz	1 + 0

We conclude this section with ways to increase microwave link capacity. Because as the operator evolves and customers surpass beyond marketing expectation, the network will grow and capacity expansion will be point of discussion in everyday operator meetings. ModCod mean modulation and coding scheme to be used while L1LA is layer one link aggregation of two or more physical carriers.

Frequency and capacity planning

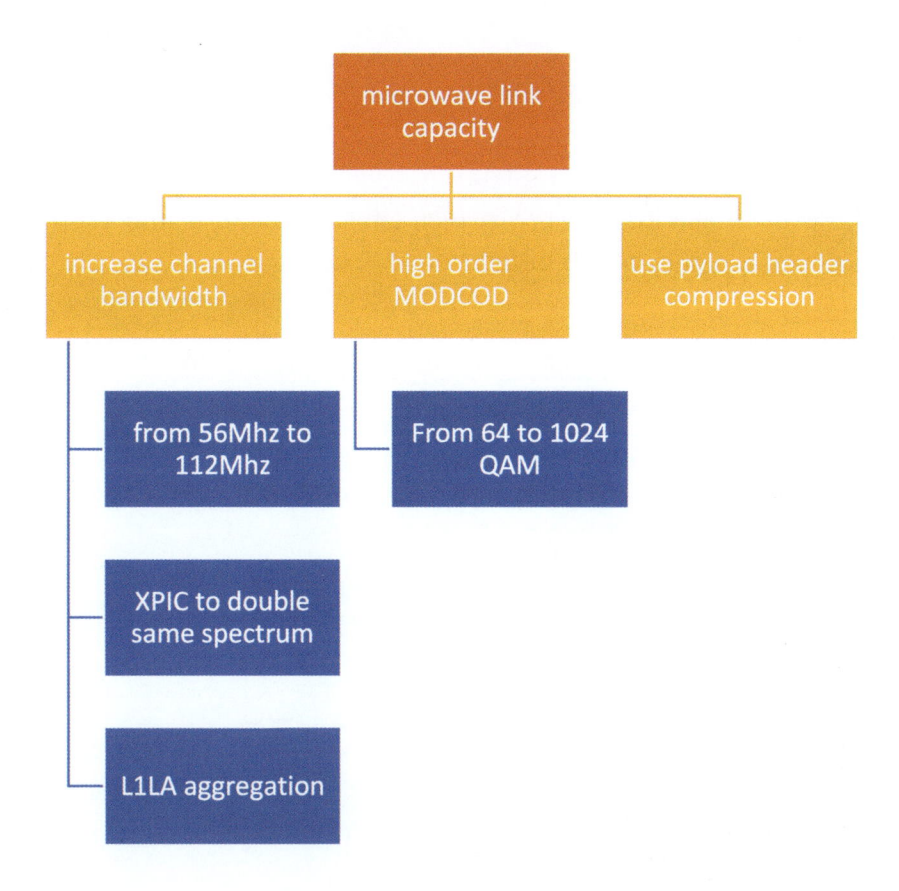

```
                    microwave link
                       capacity
        ┌──────────────────┼──────────────────┐
  increase channel     high order       use pyload header
    bandwidth           MODCOD             compression

   from 56Mhz to      From 64 to 1024
     112Mhz                QAM

   XPIC to double
   same spectrum

   L1LA aggregation
```

Figure 6.10 ways of increasing microwave transmission bandwidth

Chapter Seven

Interference analysis

Interference is one of the major causes of radio network problems that costs a lot of money and customer to the operators. Before how to combat interference is even talked about, it is important to understand what interference is and what main sources of interference are

If we transmit main signal towards receiver, then any other undesired signal that interacts with the main signal is interference. Interference can come from within the main system (intra system interference) or from an external source (inter system interference)

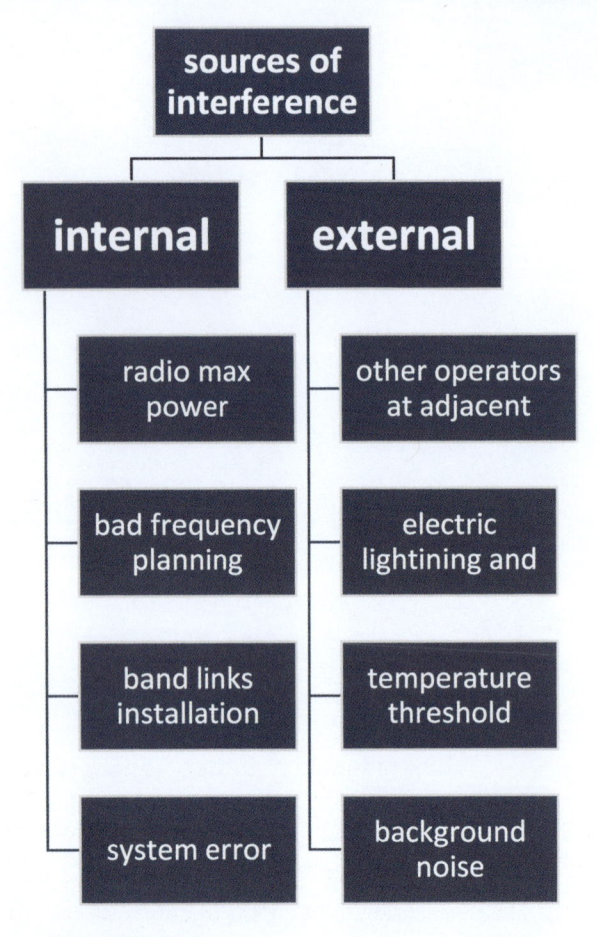

```
                    sources of
                    interference
                   /            \
              internal         external
                |                 |
           radio max       other operators
            power           at adjacent
                |                 |
          bad frequency       electric
            planning       lightining and
                |                 |
           band links        temperature
          installation       threshold
                |                 |
          system error       background
                                noise
```

Figure 7.1 Sources of interference

With some sources of microwave links interference highlighted, let us now discuss further some examples in real network interference calculation and how to successfully minimize it.

First question what is the metric we use to quantify microwave radio interference?

Let wanted carrier signal be C and interferer signal be I. Now our objective is to minimize the interferer signal strength which means C should be much higher than I so as to declare the

Interference analysis

interferer is negligible and would not have significant effect on our system. The relationship between C and I defined in logarithmic scale is shown below

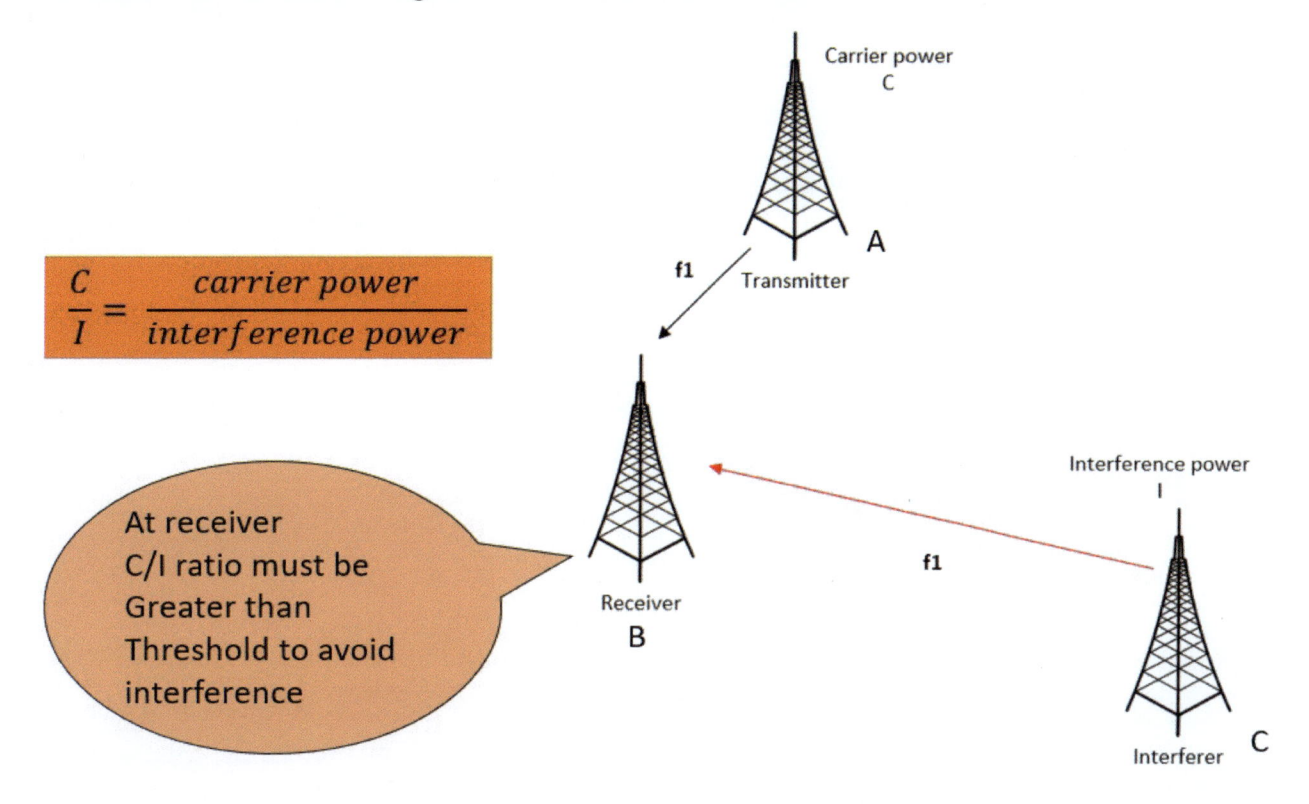

$$\frac{C}{I} = \frac{carrier\ power}{interference\ power}$$

At receiver C/I ratio must be Greater than Threshold to avoid interference

Figure 7.2 signal to interference ratio

With the ratio used to measure interference level now defined, one may ask, what is the minimum interference power that can be present at the receiver without causing degradation?

(C/I) min parameter specified in vendor data sheet is the interference threshold.

Interference analysis is done to confirm selected frequency plan for links is interference free

- First select a given band (e.g. 8GHz).
- You want re-use one channel in every hop of the backbone route
- Perform C/I analysis at every link (to verify overshoot interference)
- Perform C/I analysis at every station (to verify nodal interference)

Let us take an example of nodal interference as shown below. The link budget is calculated for the link between transmitter and receiver and that between interferer and receiver

Interference analysis

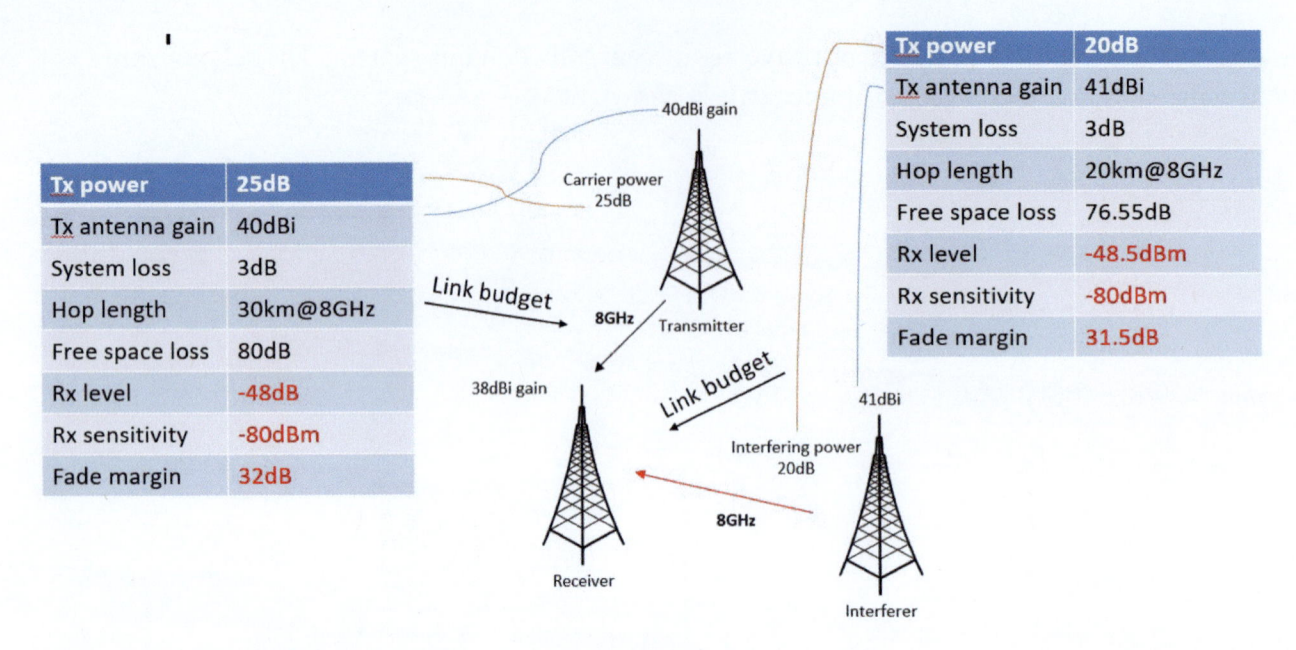

Station A (left table):

Tx power	25dB
Tx antenna gain	40dBi
System loss	3dB
Hop length	30km@8GHz
Free space loss	80dB
Rx level	-48dB
Rx sensitivity	-80dBm
Fade margin	32dB

Station C (right table):

Tx power	20dB
Tx antenna gain	41dBi
System loss	3dB
Hop length	20km@8GHz
Free space loss	76.55dB
Rx level	-48.5dBm
Rx sensitivity	-80dBm
Fade margin	31.5dB

Figure 7.3 typical link budget for interference calculation

You will have information about antenna diameter, F/B ratio, and receiver (C/I) min from equipment manufacturer data sheet. Let us assume this data for all the stations

Table 7.1 example of equipment data sheet

Antenna diameter	1.8m
Antenna F/B	50dB
(C/I)min	15dB

The C/I calculation is summarized in the following figure 7.4 for the case Station A is transmitter, B is receiver and C is interferer

116

Interference analysis

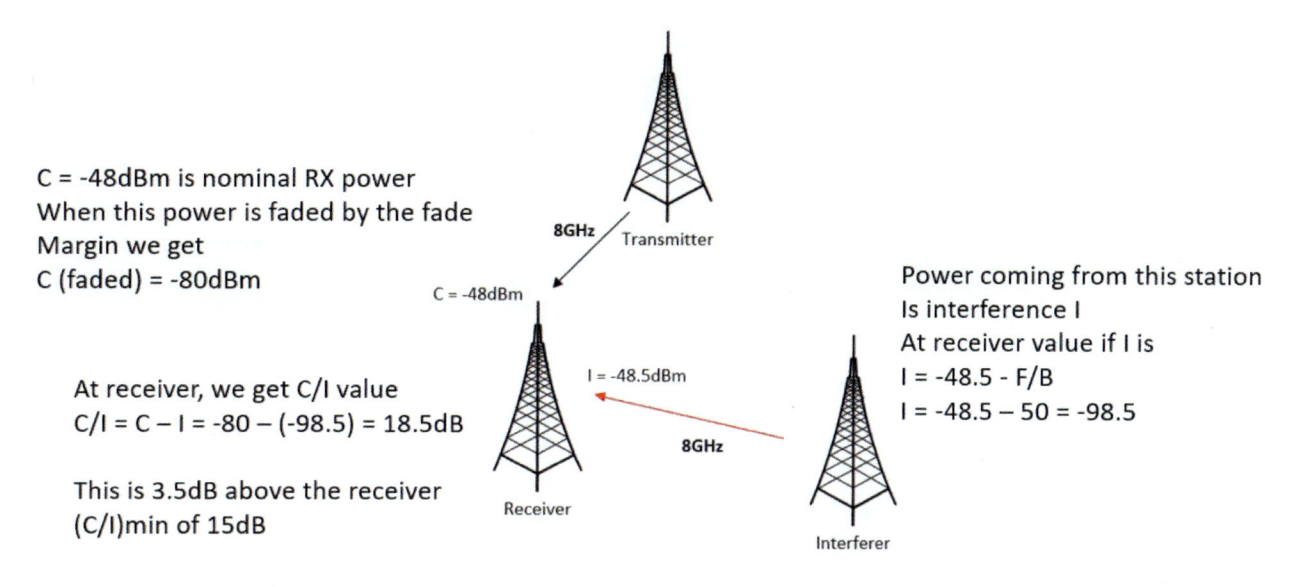

C = -48dBm is nominal RX power
When this power is faded by the fade Margin we get
C (faded) = -80dBm

Power coming from this station
Is interference I
At receiver value if I is
I = -48.5 - F/B
I = -48.5 – 50 = -98.5

At receiver, we get C/I value
C/I = C – I = -80 – (-98.5) = 18.5dB

This is 3.5dB above the receiver (C/I)min of 15dB

Figure 7.4 example of interference calculation

The C/I calculation is summarized in the figure 7.5 below for the case Station C is transmitter, B receiver, and A is interferer

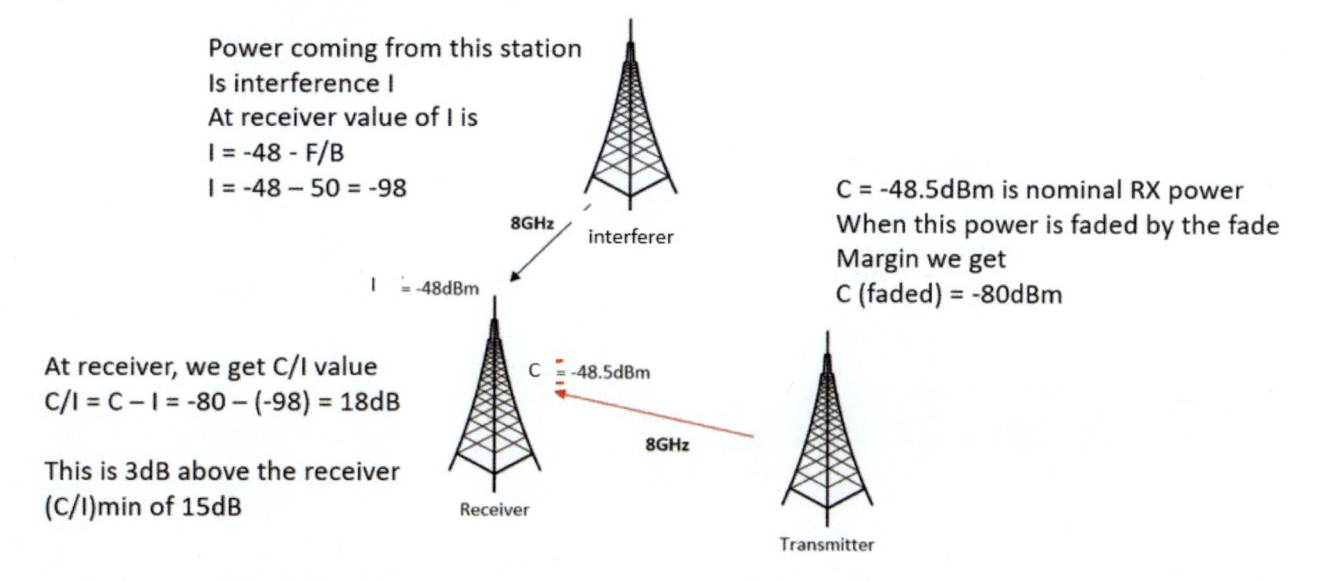

Power coming from this station
Is interference I
At receiver value of I is
I = -48 - F/B
I = -48 – 50 = -98

C = -48.5dBm is nominal RX power
When this power is faded by the fade Margin we get
C (faded) = -80dBm

At receiver, we get C/I value
C/I = C – I = -80 – (-98) = 18dB

This is 3dB above the receiver (C/I)min of 15dB

Figure 7.5 example of interference calculation

From this interference analysis using C/I ratio we can conclude that the receiver minimum C/I was not violated.

Hence same frequency can be used for the both links of the network (link A-B, link B-C)

What to do when (C/I) min is violated?

The best practice will be to employ high performance antenna.

Interference analysis

High performance antennas have larger F/B ratio.

Let us see how C/I improves when high performance with F/B of 60dB is used as illustrated below

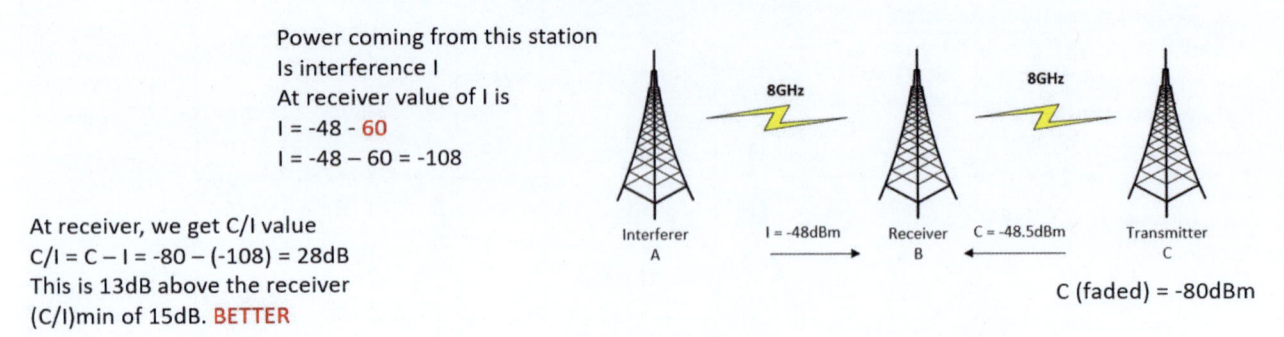

Power coming from this station
Is interference I
At receiver value of I is
I = -48 - 60
I = -48 – 60 = -108

At receiver, we get C/I value
C/I = C – I = -80 – (-108) = 28dB
This is 13dB above the receiver
(C/I)min of 15dB. BETTER

Interferer A   I = -48dBm   Receiver B   C = -48.5dBm   Transmitter C
C (faded) = -80dBm

Figure 7.6 improvement of C/I ratio when F/B of antenna is increased

This example shows using antenna of high F/B ratio of 60dB, the C/I ratio is now 28dB compared with previous case of just 18dB

Sometimes when link outage is monitored by an engineer and interference is suspected, transmitter power increase is approved to make it superior over the interferer signal. However, this does not always produced desirable results and thus not recommended especially as last resort

Increasing TX power may create noise
To other stations.
Increased power → increased coverage

Increasing power will also saturate RF
Units. System will be jammed and rebooted
Many time which reduces system lifetime

Alarm	BER degraded	Demodulator of radio B
	RSL normal	Demodulator still locked

SNR degraded at receiver B

Increase TX power

Not recommended
LAST RESORT

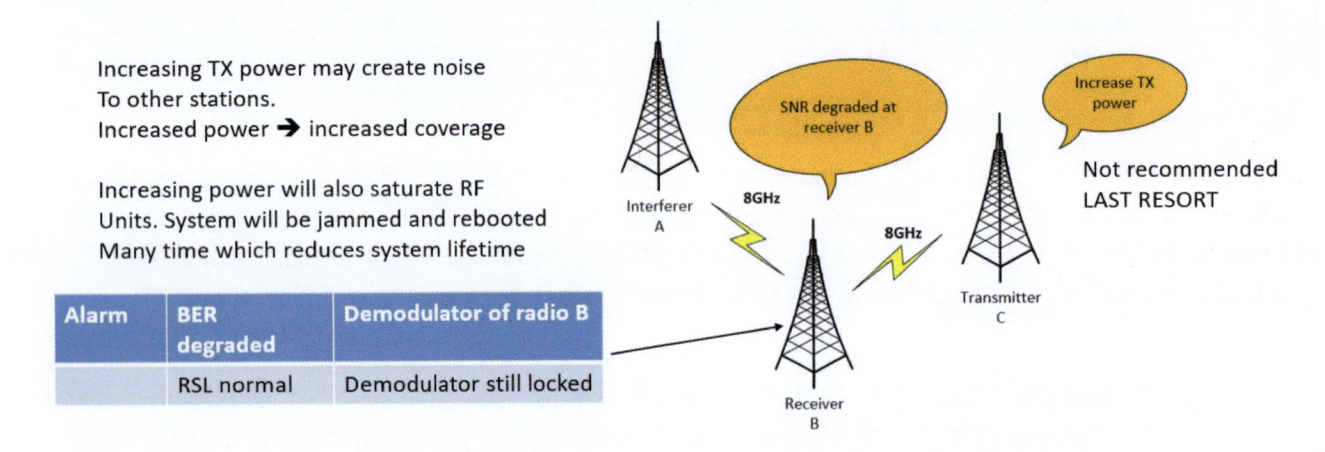

Figure 7.7 example of increases remote transmitter power to overcome the interfering signal

Finally to summarize this discussion, microwave link interference can be reduced by

- Proper spectrum scanning in the affected area using spectrum analyzer
- Reducing transmit power of the interferer site
- Tuning link frequency plan and interference analysis in Pathloss tool
- Taking advantage of the terrain to block unwanted signal
- Change frequency and see any effect on the RSL
- XPD improvement if the XPIC links interfere during rain de-polarization

Chapter Eight

Telecom regulation and spectrum monitoring

Wireless communication requires policies and guidelines that will coordinate wireless transmission from millions of stations in the world. Without this coordination, every transmission station will interfere with every other station.

For that reason, telecom has got an organ in the UN body called international telecommunication union (ITU) that is responsible for world telecommunication. The radio part called ITU-R is responsible for radio communication. There are several other international regulatory bodies such as European ETSI and American ANSI. For cellular communication standardization, we have 3GPP while for WIF standards we have IEEE

The ITU-R F-recommendations provide detailed approach on how channels for microwave radio communication are arranged for different bands. This is what many telecom operators in the world follow. Whether operators have complied with telecommunication regulation is governed by national communication authority of countries.

Chapter 6 provides detailed examples of microwave transmission frequency planning and how frequency allocation, allotment and assignment are carried out

The structure of the national telecom regulator is important before regulatory policy framework is developed. This is said because incumbent operators who hold large share of market never want effective regulation. They want to influence decision makings, service provisioning, price, and interconnectivity so that their market share is always protected against fair competition. This is against the main concept of effective telecom regulation

Thus the telecom regulatory body should be free from market actor's influence, should have independent functions monitored by the ministry of information technology, and should also have independent cash flow from spectrum and other telecom asset licenses. The cash flow also comes from penalty imposed on market aggressors.

The functions of effective regulations include

- Wireless spectral licensing and monitoring
- Price and fair competition regulation
- Interconnectivity and roaming regulation
- Environmental protection by enforcing telecom equipment compliance with health standards
- Facilitating innovative technological competition among operators
- Protecting and compensating service level agreement (SLA) between customers and operators

Let us illustrate common regulation malpractices done by incumbent operators in countries where regulation is very ineffective such as Somalia.

Case I: new telecom operator enters the market, the incumbent operator lowers the service price beyond unreasonable level such that in the short term, they can survive but the new operator has no choice but to lose significant investment and exit the market

This is very common regulatory malpractice. In this case telecomm regulation is violated especially fair price competition.

Case II: incumbent telecom operator lowers internet price by manipulating connection speed against the customer SLA contract while other operators are committed to giving customers exactly what they buy when customers does not knowledge capacity to verify that he actually got what he paid for

The customer does not have experience to test connection speed he is getting from the operator and hence enjoys lower speed but that was lower than what he paid for. Hence SLA is violated. Also competition is violated because other operators are respecting SLA and giving what the customer has bought and the aggressor operator is adverting same speed as others which is not true

Case III: a customer in remote location loses coverage of the serving operator. The coverage of other operator is available. In this case, the telecom regulation should enforce roaming policies that will allow the customer to use the available coverage of the other operator

One of the main reason telecom industry loses significant amount of money in countries where regulation is less effective is that the customer is forced to use only services of the incumbent operator as it has wider coverage than the new operators. The regulatory policy will enforce roaming to protect interests of the consumers

Case IV: distinction between cellular operator and internet service gateway provider

Many countries where telecom regulation does not exist or less effective only one company tries to be both operator and provider at the same time. This should not be allowed as it will not lead to fair competition. At the same time it does not protect economy of the country. The regulation gives recommendations on how to share common infrastructure by separating cellular operator and submarine fiber internet provider

Wireless spectrum monitoring case study

Let us take as an example of microwave spectrum licensing and monitoring. The wireless spectrum is limited and scarce resource. It is divided into different sections based on frequency. Telecom system are allocated to those sections by ITU-R

Now spectrum analyzer is the most valuable asset in dealing with spectrum monitoring. When purchasing analyzer, the operator needs to bear in mind what modules are required. Without this prior information, significant investment will be lost by buying unnecessary components

The spectrogram is one such useful component as it enables to record signal changes over fixed period of time to eliminate the error of capturing an active carrier at a time when it was down due to path failure. After scanning is finished, then carrier is up and you report as unused carrier even though it is used by certain operator

Other components required are

- Channel scanner
- Interference analyzer
- Frequency range 6 to at least 42GHz to include microwave frequency bands (table 6.2)
- External directional antenna with right frequency range and other cables

A typical setup of spectrum monitoring or operator interference analysis is shown in figure 8.1 using Anristu MS27207 analyzer. The technician climbs up tower with the directional antenna, analyzer and wireless router. The engineer will then read the output of the analyzer remotely on the ground. This example requires the technician to be highly trained on the analyzer settings. If both the engineer and the technician can climb the tower, it will produce more accurate results and hence the author recommends the latter approach. One important component missing in figure 8.1 is compass bearing that is essential to direct the antenna to all possible remote antennas. To do so the azimuth bearing database of sites should be carried to the test sites

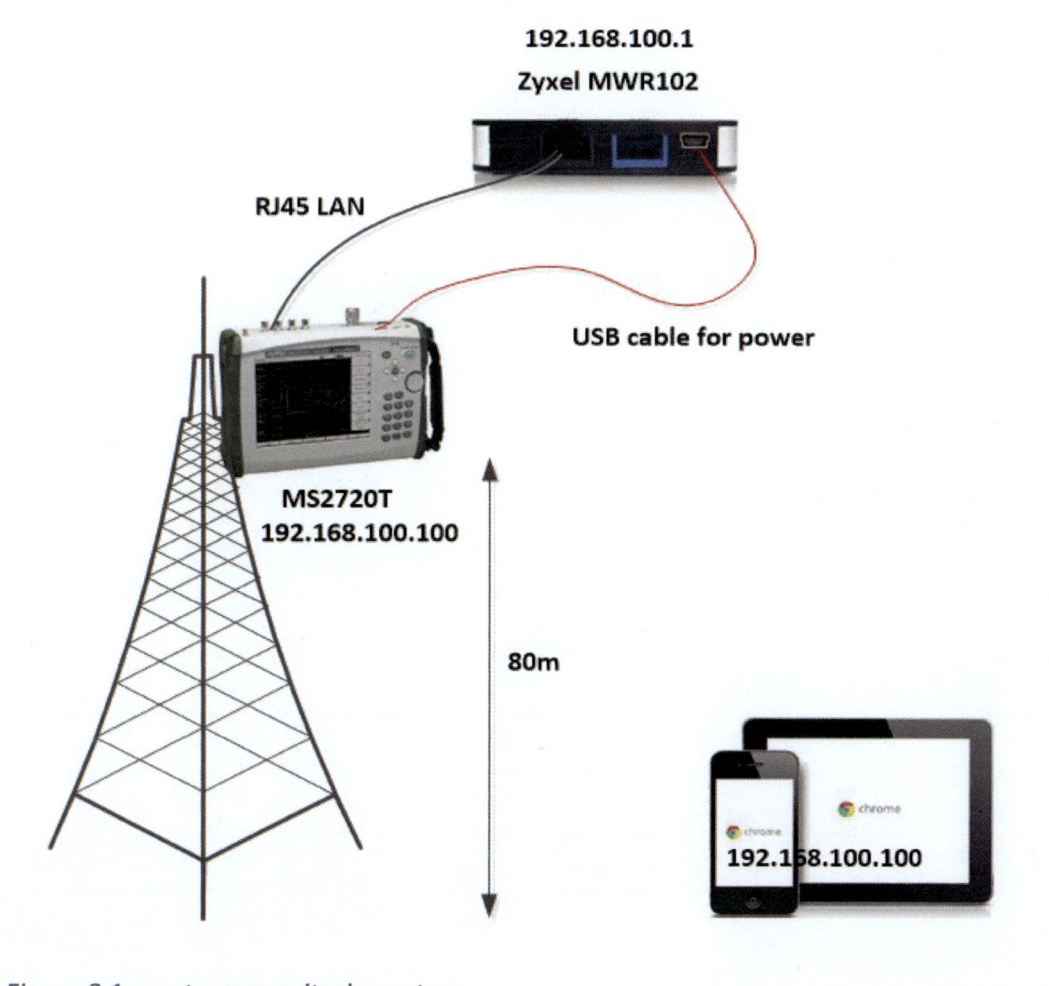

Figure 8.1 spectrum monitoring set up

From figure 8.1 we can conclude that the following requirement must be available when conducing interference analysis or spectrum monitoring

- Ms2720t spectrum master (0-32GHZ)
- Antenna (1-42GHZ)
- 1m RF cable for antenna and master interconnection
- Compass
- Rj45 cable or pocket Wi-Fi router for remote control
- Route design azimuth bearings and antenna heights file
- List of proposed channels to scan and monitor in the spectrogram

To come up with comprehensive plan before executing this task, the following methodology could be used guideline

- Specify the number of sites to scan
- Specify antenna height of each link
- Specify antenna bearing of each link
- Specify frequency band for each link
- Specify analyzer settings for each measurement
- Specify test acceptance procedures and measurement validation rules

In the link planning phase, the planning team had already established antenna height. That means where the link will be installed on the tower is known. Hence that will be the position where spectrum monitoring will be performed. An example of link azimuth bearing is shown in table 8.1

Table 8.1 example of link bearing file for spectrum monitoring

Site A	Site B	Link distance (km)	Site A bearing	Site B bearing
HQ	Birta	9.3	265	85
Golden	Star	5.4	360	170
B	I	26.7	265	70
C	J	3	Visible	Visible
D	K	32.5	324.5	144.5
E	L	29.5	7.8	187.8

F	M	12	39.16	219.16
G	N	16.5	248.6	68.6
H	O	38	111	291

And antenna height in table 8.2

Table 8.2 link antenna height for spectrum monitoring

Site A	Site B	Link distance (km)	Site A height	Site B height	Band
HQ	Birta	9.3	35	50	11
Golden	Star	5.4	35	40	11/15
B	I	26.7	40	40	7/8
C	J	3	30	30	11/15
D	K	32.5	30	30	7/8
E	L	29.5	25	40	7/8
F	M	12	20	30	11
G	N	16.5	30	21	11
H	O	38	55	75	7/8

Let us take an example of typical high site where we monitor low sub-band of the channel plan. A high site will transmit high sub-band but receive the low sub-band. So we test all channels in the low sub-band to verify if there is remote low site transmitting those low sub-bands. What is also needed is to test any nearby high sites

HIGH	LOW	BW	TR
8314	8106	56	208
8370	8162	56	208
8426	8218	56	208

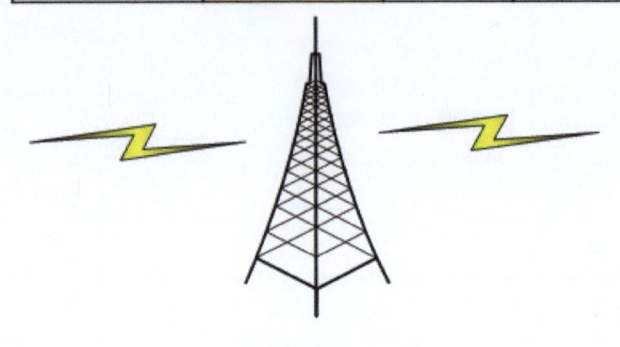

High site

Figure 8.2 a high site transmitting high sub-band will receive the low sub-band

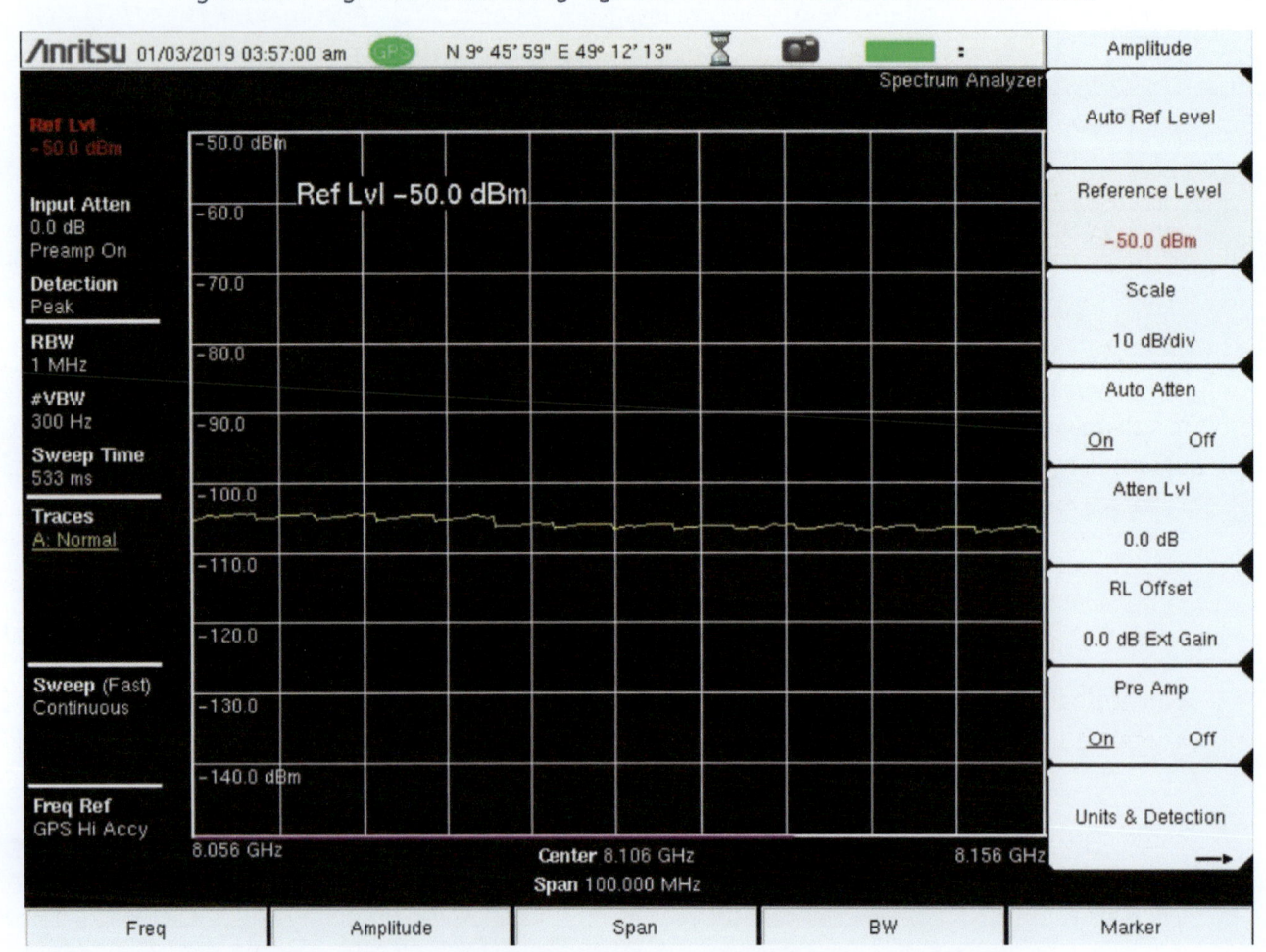

Figure 8.3 an example of observing 8GHz TR 208 channel on spectrum analyzer

From figure 8.3 above it shows that with the following spectrum analyzer settings

- Center frequency to measure is 8106MHz
- Span of 100MHz to measure 50MHz on each side of the center frequency as required bandwidth is 56MHz
- Reference signal level of -50dBm with preamplifier turned on to detect even weaker signals
- Video bandwidth of 300Hz to smoothen the trace
- Marker turned on to indicate the highest peak

That the signal under scan is clean as it is flat with no hills. The horizontal trace we see in the figure is the noise floor of the spectrum analyze receiver. If there was an interferer present, then we would have peaks somewhere on the bandwidth.

Notice also that the GPS of the site is displayed on the top part of the screen. An external GPS will be required to achieve that. It will add greater value to the final report as it validates the measurement was indeed taken on the interested site

Thus this particular channel will be suitable for this link to order radio equipment. The remaining part is to visit the low site of the link and scan the high sub-band

A second example of interference analysis or spectrum scanning and monitoring is illustrated in figure 8.4 below

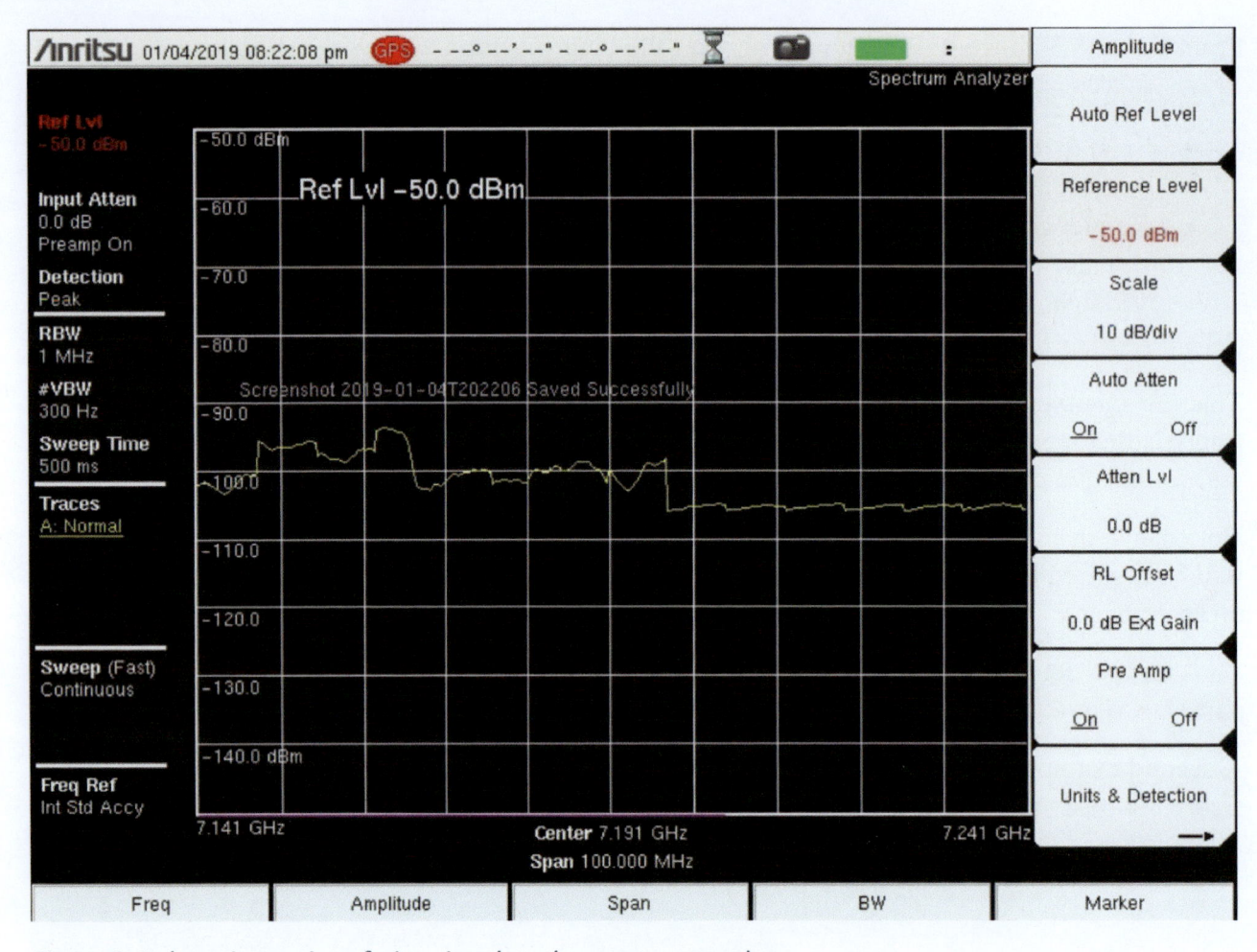

Figure 8.4 observing an interfering signal on the spectrum analyzer

As can be seen in figure 8.4 above, an interfering carrier of 7191MHz is present. The interfering signal is 15dB above the noise floor. If we use this channel and it hits our site with link margin of 10dB it will bring it down and disrupt the service. It also shows that the right of this channel is less affected than the left. This suggests not to use this channel in this particular location as it is being used by possibly other operator

Many spectrum analyzers support markers where peaks of strong signals can be marked and easily spotted. Figure 8.5 below shows the marker table right below the trace

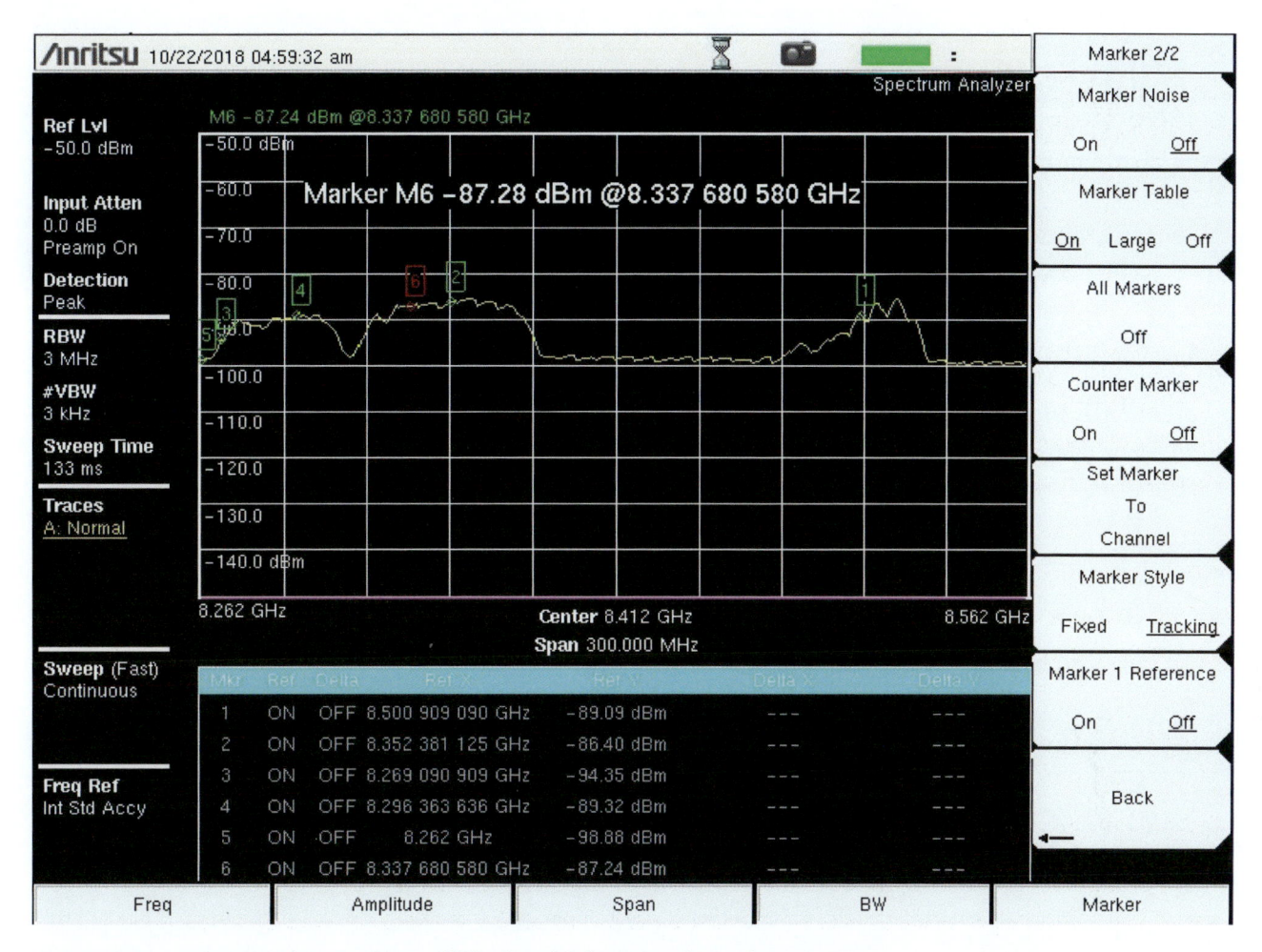

Figure 8.5 turning on the Marker will display highest peaks in the spectrum

To summarize spectrum monitoring is tedious process that also requires information management system such as proper reporting to be in place. Results of the spectrum scanning could be used to order purchase of expensive equipment or for licenses. It is therefore recommended to do it in right way with proper strategy and training

Chapter Nine

Cellular network planning fundamentals

Introduction

In the past several sections we solely focused on Microwave transmission systems that provide transport services in the frequency range 6-42 GHz. There are many newer system that also support the E-band (70/80GHz) for high capacity short haul communications

In cellular RF planning we are much concerned with the air interface between the mobile antenna and BTS antenna. The RF planner is also required to dimension certain interfaces in the core network.

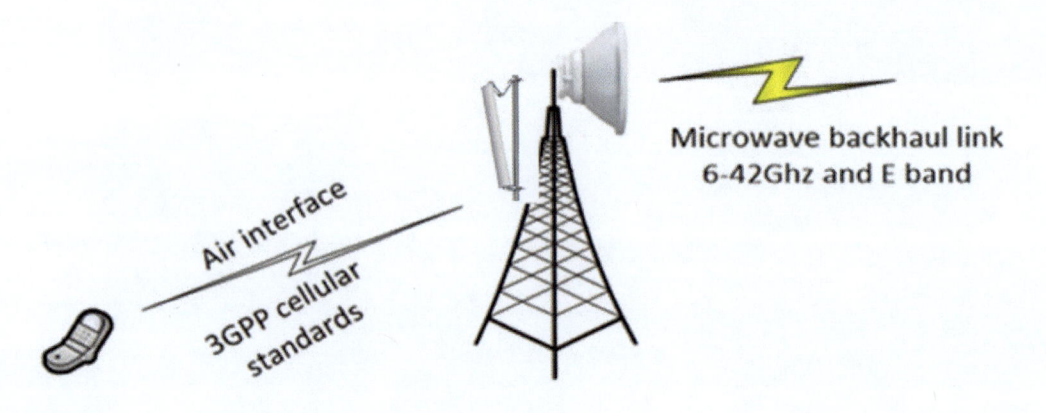

Figure 9.1 transmission link vs cellular air interface link

Cellular standards developed by 3GPP organization are widely used in mobile communications such as GSM, UMTS, LTE and the new 5G. Brief history of the evolution of these technologies is illustrated below. The need of newer technologies with higher throughput was fueled by exponential growth of subscriber density and also development of newer high bandwidth applications such as video streaming and online gaming.

As shown in the graph below, capacity increased from 9.6kbps of GSM all the way to 10Gbps for 5G practically. This has resulted in billions of mobile devices and tablets accessing the internet and many online businesses to thrive. In particular 5G standard is also promising for broader connectivity such as auto-driving cars and remote surgery.

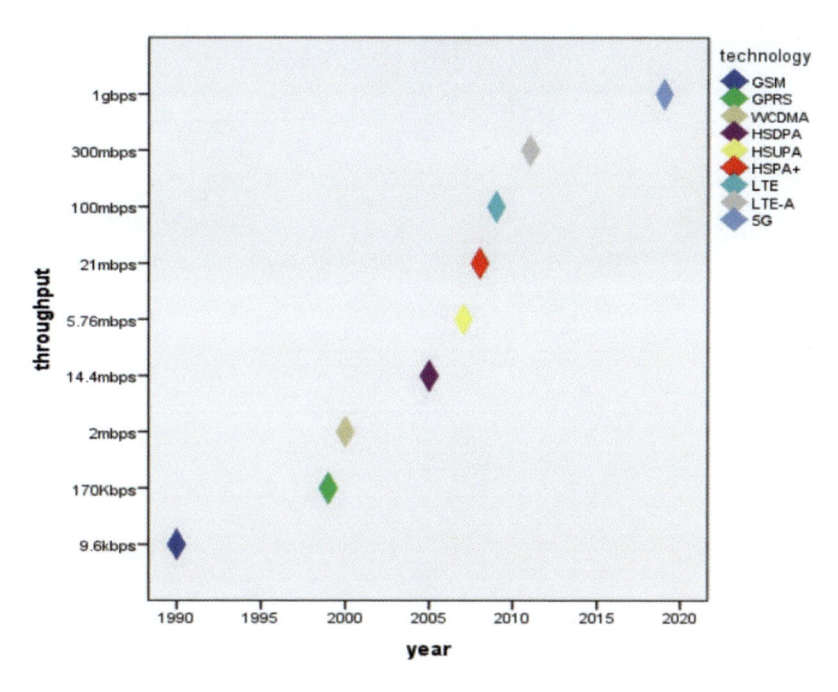

Figure 9.2 cellular technologies throughput evolution

As shown in the evolution graph above, early technologies such as GSM still dominate in the market today for low data rate voice services. Many operators in the world have now deployed 4G network while in the coming years 5G rollout will expand. This means as these cellular systems grow with capacity needs, microwave communication should also be revolutionized to compete with optical fiber.

When considering planning phase of cellular networks, the planner will set up the planning environment and execute one by one as follows

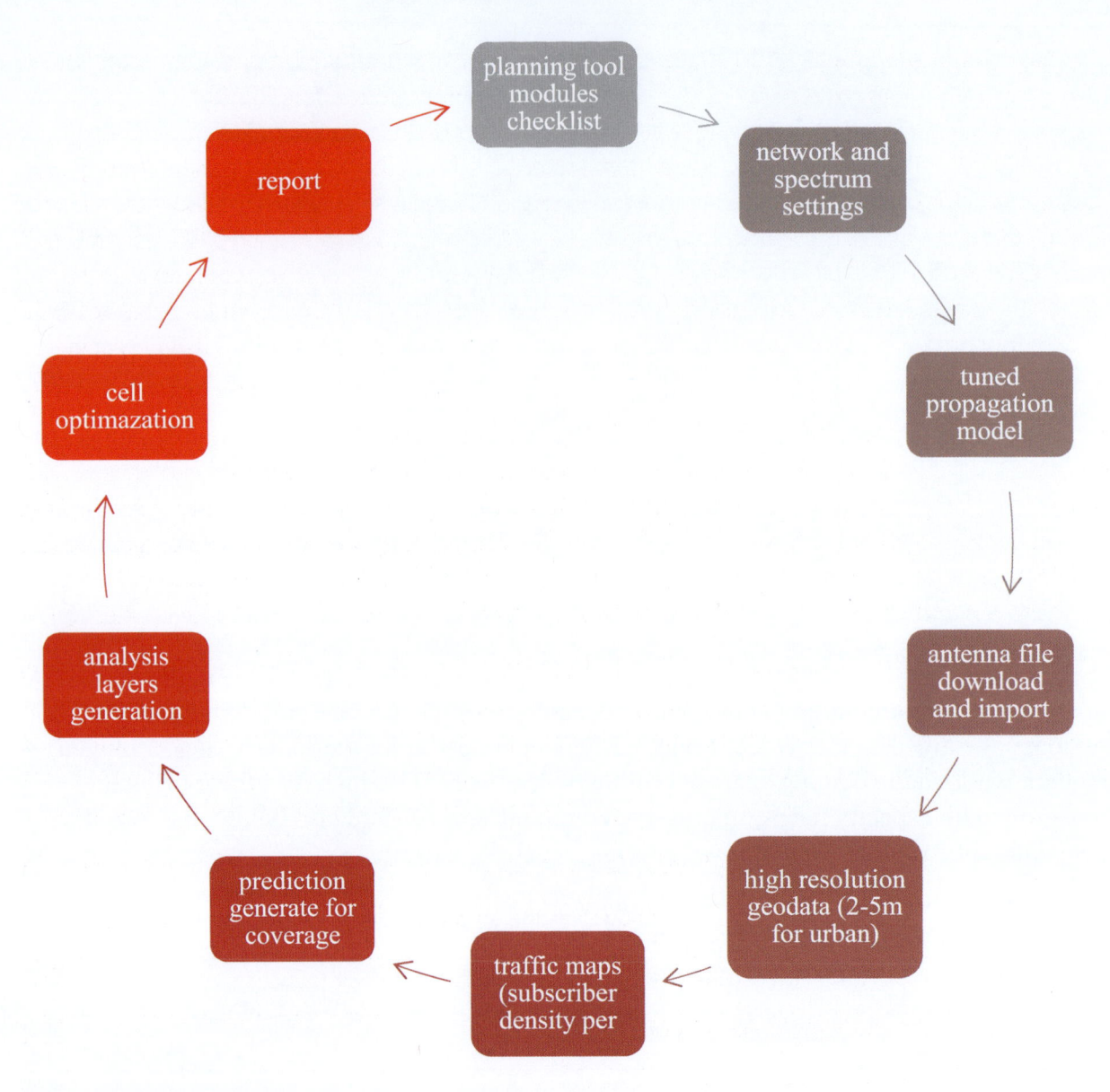

Figure 9.3 Cellular RF planning steps in planning tool

Everything starts with buying and acquiring planning tool that has modules applicable with the requirement. The planner will then set up and configure required settings in the planning environment. In the analysis layers step, various layers that relate to coverage and capacity are analyzed and visualized

Important modules to consider when purchasing expensive planning tool include

- Base modules for GSM, UMTS and LTE
- Channel scanning module
- Optimization module
- Measurement migration module

Cellular network planning fundamentals

Network settings setup is important step done by the planner before starting the design. The following table gives example of spectrum setting in GSM/UMTS/LTE systems

Table 9.1 Cellular frequency bands from GSM to LTE

Technology	UL band MHz	DL band MHz	Bandwidth MHz	Designation
GSM-900	880-915	925-960	35	B8
GSM-1800	1710-1785	1805-1880	75	B3
UMTS-2100	1920-1980	2110-2170	60	B1
LTE-2100	1920-1980	2110-2170	60	B1
LTE-1800	1710-1785	1805-1880	75	B3
LTE-800 DD	832-862	791-821	30	B20

Now let us discuss briefly basic GSM network architecture and how it overlays UMTS and LTE networks as shown below. The three technologies can co-exist in a particular operator depending on business needs and objectives. The GSM network is solely intended for voice services for customers with no smartphone. The UMTS and LTE network balance each other in data traffic. It is good practice to deploy LTE in areas where UMTS is congested to take the excess traffic away.

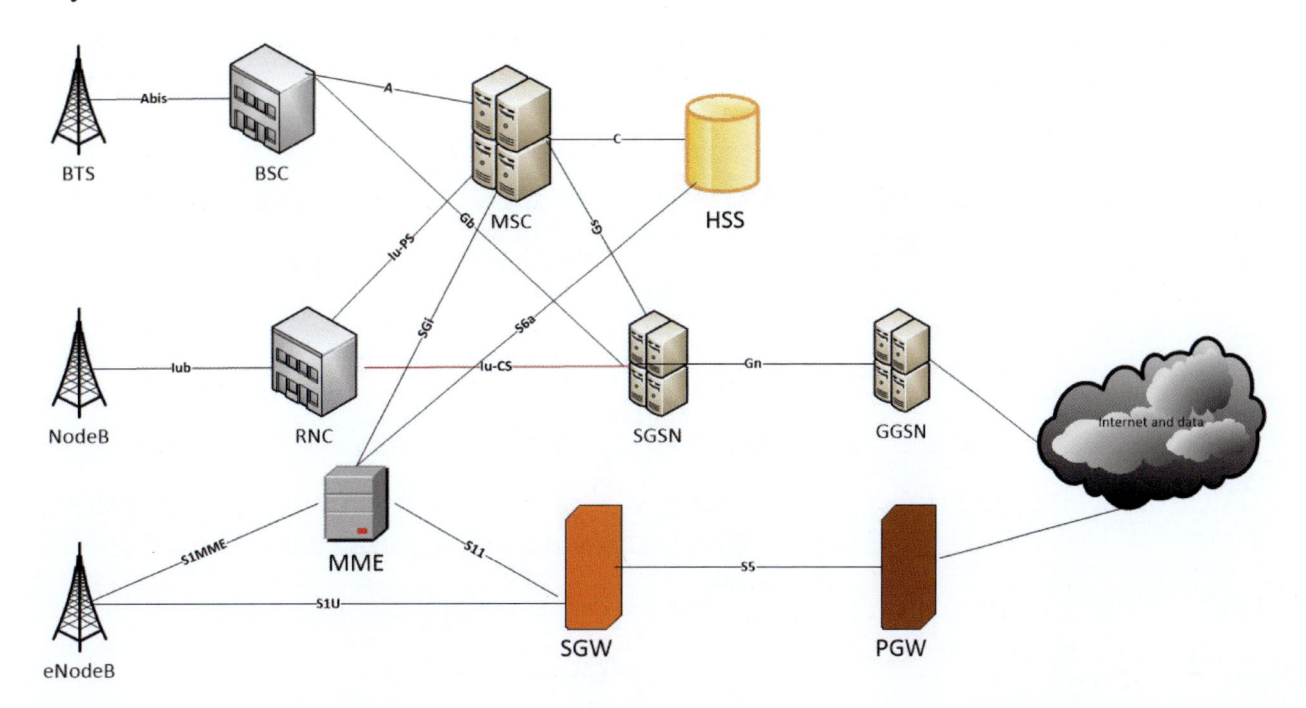

Figure 9.4 GSM UMTS LTE cellular networks components and overlay

In the upper part of the diagram the GSM network consists of BTS, BSC, MSC, and HLR (nowadays called HSS as it immigrated from TDM/SDH to IP). The user mobile terminal is connected to the BTS antenna on the cellular tower. The Abis interface between the BTS and BSC is the transmission link. It could be microwave or fiber. The function of the BSC is to control many cells serviced by many BTS in a specified location area. For call processing and signaling the BSC controller is connected to MSC where subscriber temporary database is stored. For permanent storage of the subscriber database, MSC is connected to the HLR or HSS. Later GPRS was introduced in GSM network to add low throughput data communication. The nodes SGSN and GGSN were implemented to support GPRS. These nodes were upgraded in later UMTS WCDMA network

For 3G UMTS network, the BTS is replaced by NodeB with advanced features and boards. The BSC is replaced by RNC which is the controller of all NodeB serviced by the RNC. To reduce paging and location updating overhead on the RNC, multiple NodeBs are grouped into location and routing areas and then assigned with RAC. For call processing in UMTS, the RNC is connected to the MSC or VLR and for internet connectivity the RNC interfaces with SGSN which further interfaces with GGSN directly connected to internet routers in the data center.

Finally the operator overlays LTE on the existing GSM/UMTS network. In this case eNodeB is introduced with new interfaces. The control (signaling) plane in LTE network is controlled by MME unit. LTE is an IP network, thus cannot provide voice services. As an interim solution, the voice services initiated on LTE network will directed or handed over to GSM/UMTS network in a mechanism called circuit switched fallback (CSFB)

The table below breaks down the abbreviations used in the figure above

Table 9.2 Cellular network components name abbreviations

BTS	**Base station subsystem**
BSC	Base station controller
RNC	Radio network controller
MSC	Mobile switching center
MME	Mobility management entity
HSS	Home subscriber server
SGSN	Serving GPRS support node
GGSN	Gateway GPRS support node
SGW	Serving gateway
PGW	Packet gateway
GPRS	General packet radio service

WCDMA	Wideband code division multiple access
LTE	Long term evolution
EDGE	Enhanced data rate for GSM evolution
TDMA	Time division multiple access
FDMA	Frequency division multiple access
CDMA	Code division multiple access

One key difference between GSM and later UMTS/LTE network is bandwidth required on the air interface. This difference along with others are summarized in the table below

Table 9.3 GSM UMTS LTE comparison example

	EDGE	WCDMA	LTE R8
Link bandwidth	200kHz	5 MHz	20 MHz
Theoretical throughput	384 kbps	2 Mbps	100 Mbps
Scalability		Dual carrier HSPA+ uses 10 MHz giving 42Mbps	LTE use scalable bandwidth of 1.4, 3, 5, 10, 15, 20 in R8 and carrier aggregation up to 5 giving 100 MHz in R10 LTE-Advanced giving 3Gbps in theory
Bands in Africa	900/1800	2100 and 900 re-farming	800 / 1800 / 2100
Access technique	TDMA	CDMA	TDMA
Handover	Hard	Soft	Hard
DL identifier	BSIC	PSC	PCI
Channel raster	200kHz	200kHz	100kHz
Air interface name	GERAN	UTRAN	EUTRAN
Downlink coverage	BCCH (broadcast control signal)	CPICH (pilot signal)	RS (reference signal)

Other key difference between the technologies is the different quantities measured by the phone and when doing KPI drive test. It is illustrated below

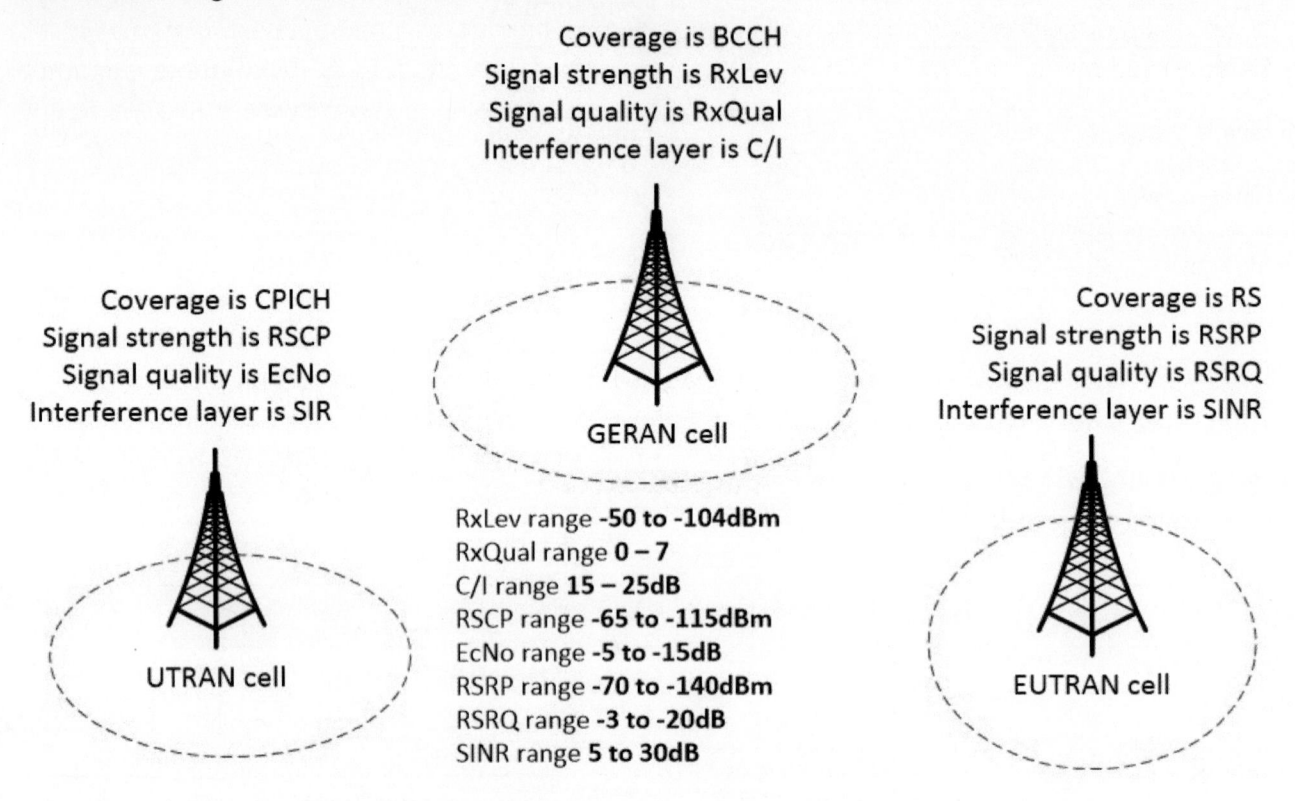

GERAN cell
Coverage is BCCH
Signal strength is RxLev
Signal quality is RxQual
Interference layer is C/I

UTRAN cell
Coverage is CPICH
Signal strength is RSCP
Signal quality is EcNo
Interference layer is SIR

EUTRAN cell
Coverage is RS
Signal strength is RSRP
Signal quality is RSRQ
Interference layer is SINR

RxLev range -50 to -104dBm
RxQual range 0 – 7
C/I range 15 – 25dB
RSCP range -65 to -115dBm
EcNo range -5 to -15dB
RSRP range -70 to -140dBm
RSRQ range -3 to -20dB
SINR range 5 to 30dB

Figure 9.5 GSM UMTS LTE RF coverage comparison

The ranges selected depends on operator's objective and design requirement. Usually the more positive value the better. In macro cell station, the signal strength at cell edge is very weak and it where coverage holes arise during network planning and launch.

Common things that affect coverage of all those technologies include

- Antenna height
- Antenna tilt (electrical and mechanical)
- Transmit power
- Cell selection threshold
- Load
- Imbalance between uplink and downlink

Cellular operators' tough choices … Do they need to deactivate 3G or 2G to boost 4G and 5G NR Non-standalone

Traditionally it was cost-effective for operators to integrate 4G solution into existing 2G/3G infrastructure without buying expensive IMS VOLTE platform. In the early days of 4G rollout, spectrum licensing was not quite easy as 4G demanded large bandwidth of 20MHz to deliver the high capacity it promised. In certain hotspot areas, HSPA+ traffic was peaking and LTE was required to take the excess traffic away. Soon massive LTE rollout was necessary.

Using Marketing tools, operators are coming to this conclusion. Data traffic demand is exponentially growing while circuit-switched telephone is declining in urban centers

| Deploying R10/R11 LTE is getting hard because of spectrum shortage |
| Large number of component carriers are required to achieve carrier aggregation |
| Customers in urban centers need 4G availability all the time which makes HSPA+ |
| In rural areas where operators get large revenues, Circuit switched telephone is |

Engineers know that LTE is IP-network while 2G/3G support circuit switched calls. This means still 2G/3G are required to make calls from LTE phone in an interim process called circuit switched fallback (CSFB). By the time fully blown IMS system is put in the network, CSFB will go away

Many operators in the world are deactivating their 2G or 3G systems to make room for future data demand explosion that will require extensive presence of 4G and the new 5G. 5G will bring many new things in reality like machine to machine communication, remote surgery, auto-driving cars, enhanced multimedia ultra HD applications and massive MIMO support.

The better approach would to be to deactivate 3G network and leave 2G. Many traditional phones are still highly available and a lot of people still use both in urban and rural. 3G systems experience massive cell breathing and pilot pollution. In Africa 3G systems occupy B1 which could be refarmed into LTE.

In the end it would be for operator's business strategy and long term vision that will decide this course. Market penetration of smartphone capable of 5G will likely increase in the near future and a lot of customer won't be happy with the current setting especially with the introduction of ultra-high speed application that require ultra-low latency

GSM

GSM is one of the older cellular communication still used today. Millions of only GSM capable phones are still use in today in many countries. It is mainly used for voice services. The network components for GSM cellular system for only mobile to mobile local calls is shown below along with basic call processing

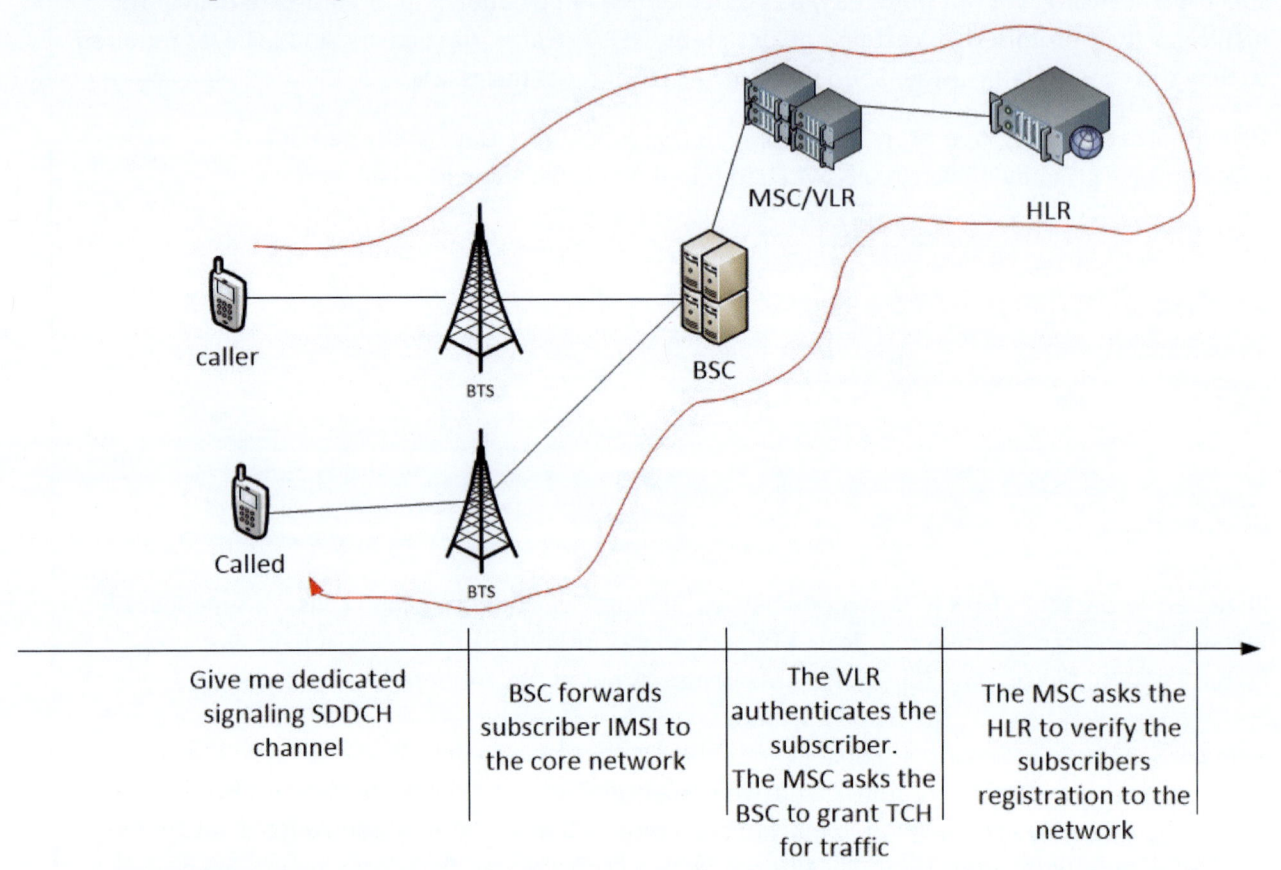

Figure 9.6 Basic GSM call flow

Every phone in the GSM cell will listen to BCCH that carries SIB messages to be used in idle mode cell selection and reselection. If two adjacent GSM cells were incorrectly assigned same BCCH ARFCN, then the cell identifier BSIC is used to differentiate those two cells.

GSM BSIC is made up of network color code (NCC) and base station color code (BCC). It has 1 – 63 range

BSIC = NCC + BCC where both NCC and BCC take 0 – 7 range

As an example of GSM frequency planning, consider an operator which licensed 8 MHz in B8 spectrum from the telecommunication authority. Remembering that the link between the mobile terminal and the BTS antenna is bi-directional. Hence the operator will have 8 MHz in both UL and DL.

Suppose that the spectrum block given to the operator UL is 907 – 915 and DL is 952 – 960 as shown below

Table 9.4 8MHz GSM ARFCN table

85	952	907
86	952.2	907.2
87	952.4	907.4
88	952.6	907.6
89	952.8	907.8
90	953	908
91	953.2	908.2
92	953.4	908.4
93	953.6	908.6
94	953.8	908.8
95	954	909
96	954.2	909.2
97	954.4	909.4
98	954.6	909.6
99	954.8	909.8
100	955	910
101	955.2	910.2
102	955.4	910.4
103	955.6	910.6
104	955.8	910.8
105	956	911
106	956.2	911.2
107	956.4	911.4
108	956.6	911.6
109	956.8	911.8
110	957	912
111	957.2	912.2
112	957.4	912.4
113	957.6	912.6
114	957.8	912.8
115	958	913
116	958.2	913.2
117	958.4	913.4
118	958.6	913.6
119	958.8	913.8
120	959	914

8 MHz spectrum

40 channels

ARFCN table

121	959.2	914.2
122	959.4	914.4
123	959.6	914.6
124	959.8	914.8

To avoid adjacent channel interference with neighboring operators, 0.2 MHz will be cut from the edges of the block. Hence practically the operator will have UL as 907.2 – 914.8 MHz and DL as 952.2 – 959.8 MHz

How many channels or TRX are available from the assigned spectrum of 8 MHz?

To answer this question the GSM standard specifies GSM radio frame or TRX as 200 kHz (0.2 MHz) of bandwidth

$$\#channels = \frac{spectrum\ block}{spectrum\ raster} = \frac{8}{0.2} = 40$$

The operator will then have 40 channels. In GSM terms, this is called 40 ARFCN (absolute radio frequency channel number)

In this way we have divided the available 8 MHz bandwidth into 40 channels or ARFCN. This is called frequency division multiplexing (FDMA)

GSM also uses time division multiple access (TDMA) to further subdivide each ARFCN into 8 time slots. Each user can communicate on 1 time slot. Not all timeslots will carry user data. Some will be reserved for signaling overhead. For example with 8 time slots in single ARFCN, the first time slot will carry signaling channel called BCCH, while the remaining 7 time slots will carry customer data. The mobile will first read and decode the BCCH before it reads the data or TCH (traffic channel)

$1\ ARFCN = 8\ time\ slots$

$40\ ARFCN = 40\ x\ 8 = 320\ time\ slots$

This means 320 users can simultaneously transmit and receive on the system at any given time. In GSM, there is also another concept of frequency re-use as spectrum is not enough to be used in all cells. This means an ARFCN used in particular cell or sector will be re-used in another cell separated at reasonable distance from the other cell.

Thus if one ARFCN is re-used in N times, total capacity of the system will be 40 times N. This in turn means, the system will handle 320N users at any given time. The ARFCN parameter is also called TRX in the BTS hardware.

The next question is how to configure those TRXs?

One thing to remember is that urban centers need more capacity and hence more TRX than rural area.

As an example of common configuration in urban centers is called 4 x 3 plan. This plan means 4 TRXs to be assigned to each sector of a cell that has three sectors each covering 120 degrees of the coverage. Thus the total cell capacity will be 12 TRX. Now we have use 12 out of the 40 TRX license we purchased. The 4 TRX assigned to each sector comprise of 1 TRX for BCCH plus SDDCH and 3 TRX for TCH as shown below

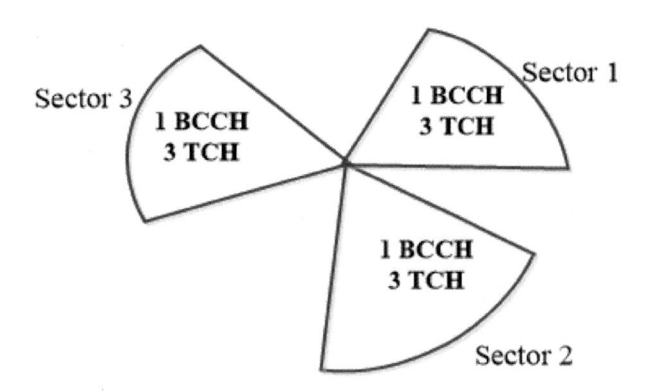

Figure 9.7 GSM example ARFCN 4 x 3 configuration

So far we have seen assigning 12 TRX to a particular cell of 3 sectors. But we only have 40 TRXs which are very small. Therefore we have to group our entire network cells into clusters. The first few clusters will consume the entire bandwidth, then we will re-use the same bandwidth again for other cells that are far away to avoid co-channel interference.

Practical cluster size used are 3, 7 and 11. In the 3 cluster size, we will assign the entire bandwidth to 3 adjacent sites. This scheme results in large bandwidth for each cell but co-channel and adjacent-channel interference increases.

In practical network design, the RF planning tool will do GMS frequency planning automatically. It will however need

- Coverage prediction to be generated
- Interference matrix for each best server to be generated
- Neighbor plan to be generated

For these algorithms to produce correct result, a high resolution GIS database and tuned propagation model is recommended. This adds to the cost of the project.

Before we go into other technologies, what happens GSM phone is first powered on?

It goes through the following initial procedure

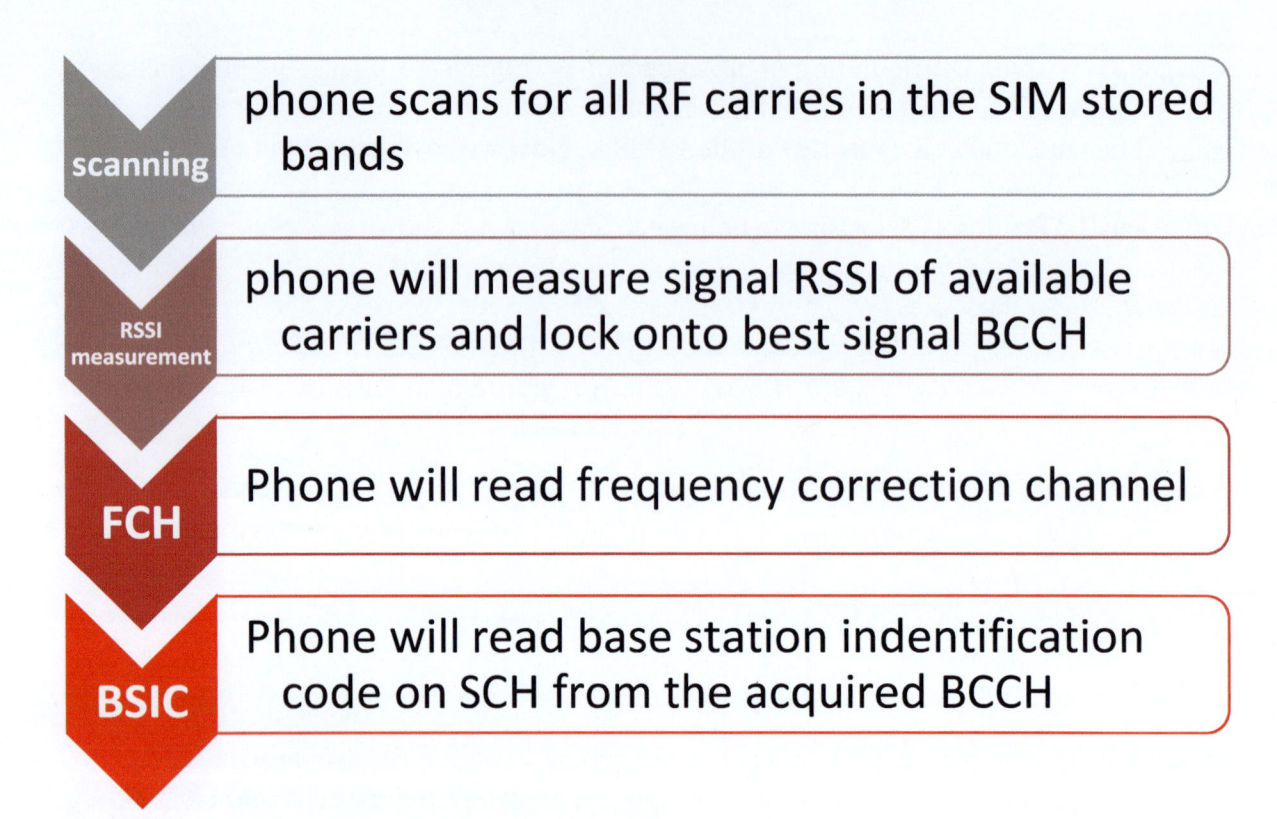

scanning: phone scans for all RF carries in the SIM stored bands

RSSI measurement: phone will measure signal RSSI of available carriers and lock onto best signal BCCH

FCH: Phone will read frequency correction channel

BSIC: Phone will read base station indentification code on SCH from the acquired BCCH

Figure 9.8 GSM basic process after phone power on

When planning GSM cells or network the following considerations must be met

- The co-channel interference layer metric C/I must be above 25dB for 95% of the coverage area as shown below. The C/I mean logarithmic ratio between carrier signal C and co-channel interfering signal. This ratio should be high enough. This could be achieved with good frequency planning as well appropriate site settings such as antenna height and antenna tilt.

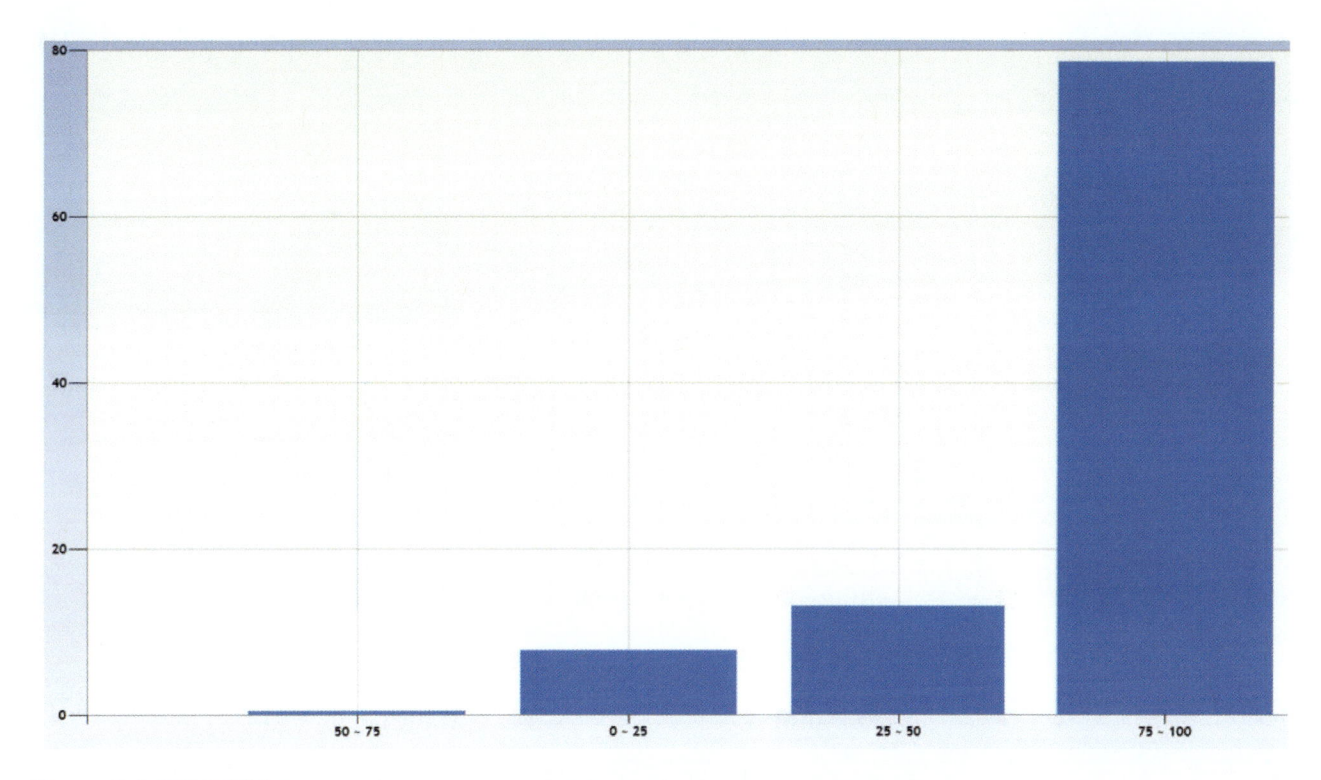

Figure 9.9 GSM C/I layer

- The RxQual which quantifies signal quality should be greater than 4 with best being 0. In frequency hopping systems, RxQual sub is used when performing drive testing KPI evaluation

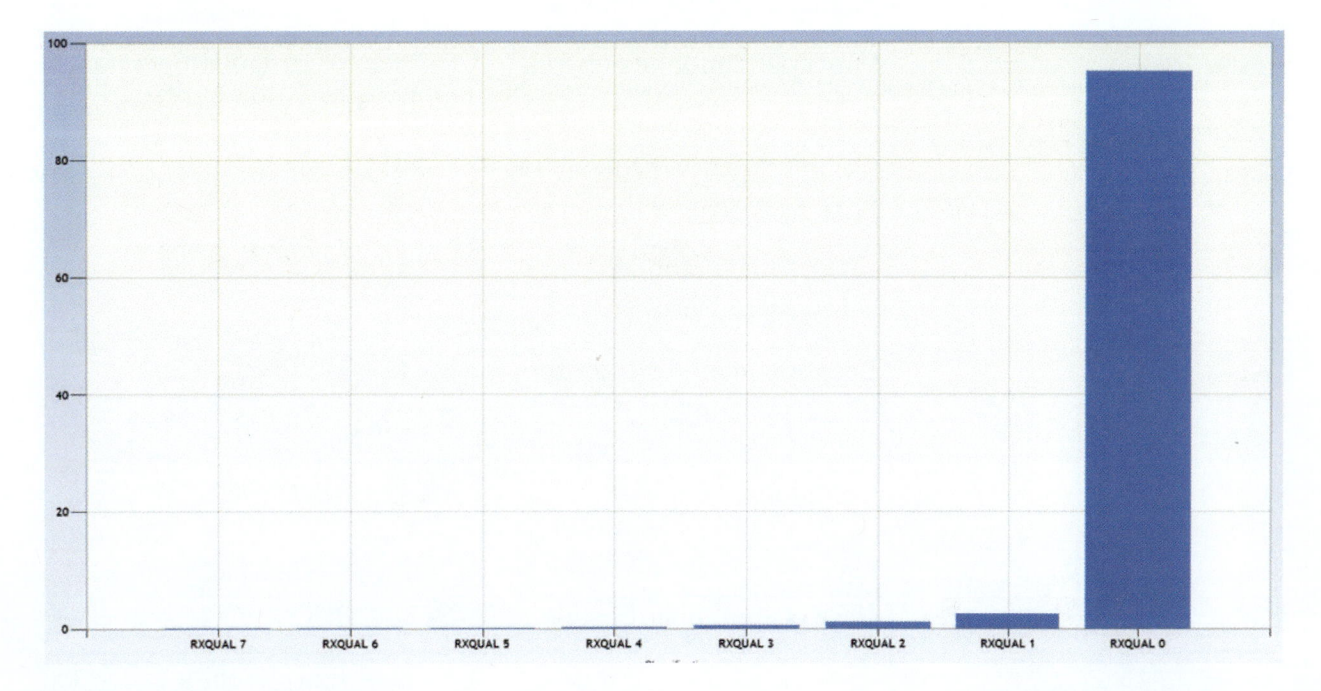

Figure 9.10 GSM RxQual layer

Also when considering coverage planning in the RF tool, the signal strength should be above -80dBm for 95% of urban areas as shown below

>=Value	<Value	Count		Color
-105.01	-105	0	(0 %)	
-105	-95.01	160	(0.05 %)	
-95.01	-95	0	(0 %)	
-95	-85.01001	6620	(2 %)	
-85.01001	-84.99999	26	(0.01 %)	
-84.99999	-75.01	88369	(26.75 %)	
-75.01	-75	189	(0.06 %)	
-75	-60.19248	183639	(55.59 %)	

Figure 9.11 Example of GSM RSSI coverage

Some RF identifiers used in planning and drive test phases

Table 9.5 some import GSM RF parameters

Parameter	Definition	Range
MNC	Mobile network code unique for the particular operator	2 digits
MCC	Mobile country code unique for particular country. Example 252 for Somalia	3 digits
PLMN-ID	Public land mobile network identifier	Unique to particular operator in the country
CI	Cell identifier to separate the three sectors of cell	1 – 65535
IMSI	International mobile subscriber identifier	MCC + MCC + MSIN
LAI	Location area identifier for phone updating	MCC + MNC + LAC 1 - 65535
CGI	Cell global identifier	MCC + MNC + CI 1 – 65535
BSIC	Base station identification code	NCC + BCC NCC = network color code BCC = base station color

		code

Shown in table 9.6 below is an example of GSM RF template sheet typically developed by RF planner using planning tool such as Mentum Planet. The GSM cell will be configured on this parameters

Table 9.6 example of GSM cell RF configuration parameters

Site name	Sector ID	CI	LAC	BSIC	BCCH	MA list	Hopping	MAIO	HSN	Azimuth	Tilt
Golden01	1	10011	1	41	1000		RF	0,1,2	0	0	3
Golden02	2	10012	1	42	1003		RF		0	120	2
Golden03	3	10013	1	43	1005		RF		0	360	0

If we shed some light on the table above, we can say the following

- Three sectors of the same site named by Golden
- Cell ID (CI) defined for each cell to uniquely identify within the operator PLMN
- Location area code (LAC) assigned to a group of neighboring cells to track phone location and update whenever the phone moves from one location to another. Also used to page phones in the same location for incoming call. Bad LAC planning can overuse system resources
- Each sector assigned with base station color code and BCCH downlink carrier
- Each assigned with mobile allocation (MA) list of TCH AFRCNs to hop from one to another to reduce co-channel and adjacent channel interference. The ARFCN to start the hopping from is specified by the parameter MAIO (mobile allocation index offset). The pattern to hop over the MA list is specified by the parameter HSN (hopping sequence number) with value 0 indicating cyclic hopping

UMTS

UMTS network has also evolved over the years to meet with increasing capacity needs as shown in the below timeline

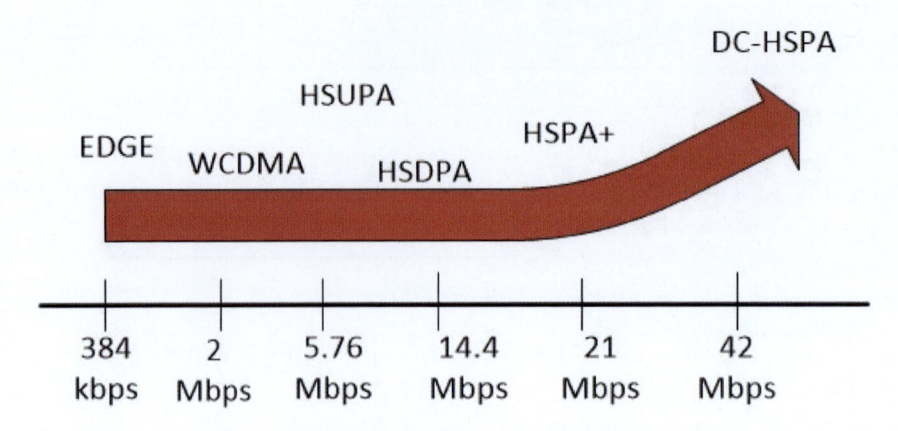

Figure 9.12 UMTS throughput evolution

Dual carrier - high speed packet access (DC-HSPA+) is an enhancement of H+ as it uses two 5 MHz dual carrier aggregation.

A phone camped on HSPA network will be in two states just like other technologies. In idle mode the phone is not sending or receiving any data, it is just locked on the cell and receiving broadcast SIB messages. It will do cell reselection to other 3G cells or inter-RAT reselection to more suitable GSM/LTE cell. In connected mode the phone is in communication. The phone will be sending measurement reports to the nodeB to initiate handover.

Some of the SIB messages in idle mode broadcasted to the WDCMA cell are SIB3 for serving cell information, SIB11 for neighboring cells and SIB19 for LTE redirection after completion of CSFB call

Now let us talk about frequency and code planning in UMTS networks. UMTS networks are deployed on wideband 5 MHz bandwidth. The center frequency is calculated from UARFCN in the frequency band. As an example suppose that 5 MHz was licensed by a new operator. Their planning team installed 50 sites in the main city. With 50 sites, total of 150 cells are available. In UMTS all those cells are given the 5 MHz carrier. The question is how to prevent interference if all cells are assigned with the same carrier. The answer is UMTS uses codes called primary scrambling codes (PSC) to differentiate between different cells. It is the job of the RF planner to plan and assign PSC to cells.

The prerequisite for PSC planning in the planning tool is neighbor planning. The neighbor planning will automatically produce best servers and all possible cells that are adjacent to it. It can be based on probabilities or interference matrix. The PSC codes are used in the downlink and 512 of such codes are available. In the uplink the RNC automatically assigns the codes and hence the RF planner is not required to plan PSC for the uplink

The PSC plan assigned to example cells is shown below. If the cells are many then PSC re-use should be handled carefully. The main point is that cells in the active set should not have the same PSC. Also neighbors of the active set should not use the same PSC plan.

```
        337
         ▲
        / \
       ↙   ↘
     338   342


                        286
                         ▲
                        / \
                       ↙   ↘
      357            285   368
       ▲
      / \
     ↙   ↘
   310   311
```

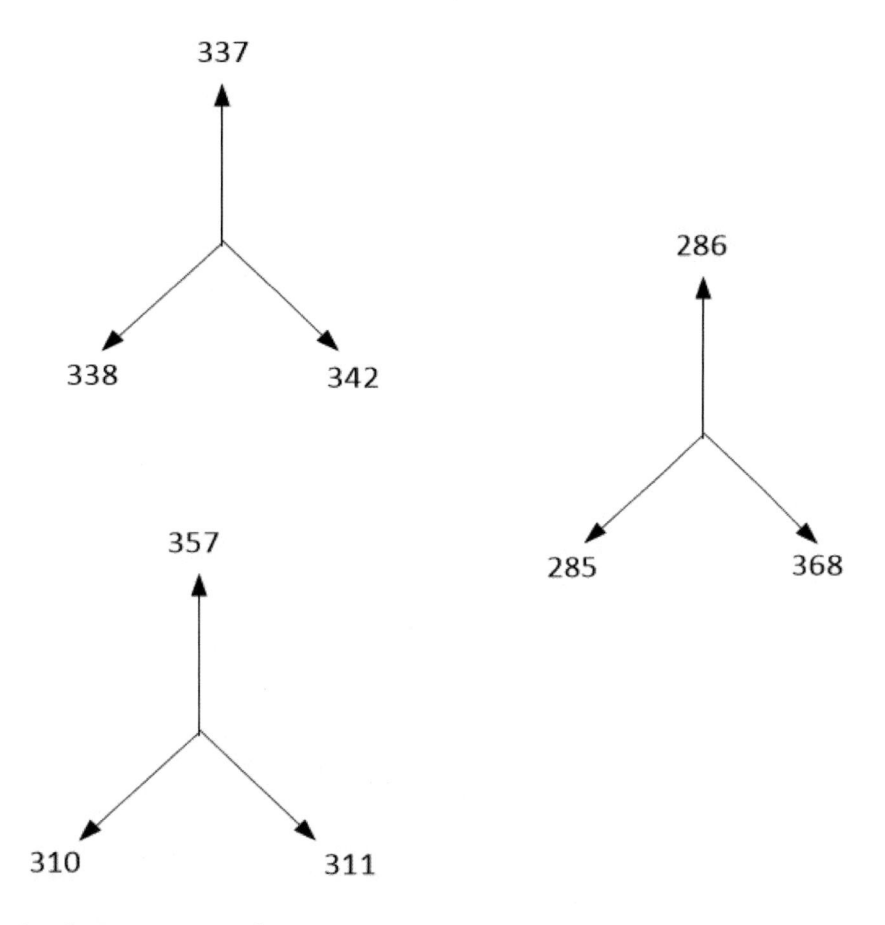

Figure 9.13 Example of WCDMA PSC plan

The RF planner will then apply the planned PSC to cells as well as neighbors list plan. As we shall see in later part of this chapter, performance issues of new network deployments are caused by missing neighbors in bad neighbor plan.

In Summary, performance issues in newly deployed HSPA network may arise from a number of factors including

- Missing neighbors due to bad neighbor plan
- Overshooting cells not included in the active set
- Installation issues such as VSWR out of range
- Incorrect site settings such as excessive tilt or azimuth away from focus area
- Incorrect parameter configurations such as cell reselection, handover and power control
- Site power disruption or fault voltage

In the upcoming discussion we will take a look at these issues deeply and suggest ways to overcome them.

An example of intra-RAT (only other UMTS cells) neighbor plan table generated in the planning tool is shown below

Table 9.7 example of UMTS neighbor plan RF template

Server Technology	Neighbor Technology	Server Site ID	Server Sector ID	Server Cell ID	Neighbor Site ID	Neighbor Sector ID	Neighbor Cell ID	Priority
WCDMA	WCDMA	Cell1090	4	21000	Cell101	6	21001	0
WCDMA	WCDMA	Cell1090	4	21000	Cell102	6	21002	5
WCDMA	WCDMA	Cell1090	4	21000	Cell103	4	21003	2
WCDMA	WCDMA	Cell1090	4	21000	Cell104	5	21004	7
WCDMA	WCDMA	Cell1090	4	21000	Cell1012	4	21005	1
WCDMA	WCDMA	Cell1090	4	21000	Cell107	6	21006	3
WCDMA	WCDMA	Cell1090	4	21000	Cell109	5	21006	6
WCDMA	WCDMA	Cell1090	4	21000	Cell1022	5	21007	4
WCDMA	WCDMA	Cell1090	4	21000	Cell1020	5	21008	8
WCDMA	WCDMA	Cell1090	4	21000	Cell1015	4	21009	10
WCDMA	WCDMA	Cell1090	4	21000	Cell1017	4	210010	9
WCDMA	WCDMA	Cell1090	4	21000	Cell102	4	210011	12
WCDMA	WCDMA	Cell1090	4	21000	Cell1021	5	210012	14
WCDMA	WCDMA	Cell1090	4	21000	Cell1023	6	210013	22
WCDMA	WCDMA	Cell1090	4	21000	Cell1026	5	210014	15
WCDMA	WCDMA	Cell1090	4	21000	Cell1027	7	210015	11
WCDMA	WCDMA	Cell1090	4	21000	Cell1028	6	210016	21
WCDMA	WCDMA	Cell1090	4	21000	Cell1030	6	210017	17
WCDMA	WCDMA	Cell1090	4	21000	Cell102	9	210018	16
WCDMA	WCDMA	Cell1090	4	21000	Cell1031	4	210019	13
WCDMA	WCDMA	Cell1090	4	21000	Cell102	8	210020	20

Earlier we said the operator licensed 5 MHz of UMTS spectrum in B1. Assume the spectrum is 2130 – 2135. This 5 MHz is not all usable. Edges will be left off to separate from adjacent operators. The RF planner will configure 10650 UARFCN on the cells and RNC. The sub-channel raster is 200 kHz. Hence 25 such channels are required to get center frequency of 5 MHz bandwidth.

Table 9.8 5MHz UMTS carrier UARFCN

UL UARFCN	UL carrier	DL UARFCN	DL carrier
9700	1940	10650	2130
9701	1940.2	10651	2130.2
9702	1940.4	10652	2130.4
9703	1940.6	10653	2130.6
9704	1940.8	10654	2130.8
9705	1941	10655	2131
9706	1941.2	10656	2131.2
9707	1941.4	10657	2131.4
9708	1941.6	10658	2131.6
9709	1941.8	10659	2131.8
9710	1942	10660	2132
9711	1942.2	10661	2132.2
9712	1942.4	10662	2132.4
9713	1942.6	10663	2132.6
9714	1942.8	10664	2132.8
9715	1943	10665	2133
9716	1943.2	10666	2133.2
9717	1943.4	10667	2133.4
9718	1943.6	10668	2133.6
9719	1943.8	10669	2133.8
9720	1944	10670	2134
9721	1944.2	10671	2134.2
9722	1944.4	10672	2134.4
9723	1944.6	10673	2134.6
9724	1944.8	10674	2134.8
9725	1945	10675	2135

One important concept to bear in mind is WCDMA uplink power control when doing cell planning. The UE has to access the cell with specific uplink power in open loop power control during random access (RACH). When it accesses the network and acquires user plane connection, the UE needs to monitor transmit power based on its distance from cell and transmit power commands (TPC) from the nodeB. It uses power control mechanism called inner loop power control and the frequency it happens is 1500Hz. Remember WCDMA is an interference based network and it is similar to 100 people in room each speaking loudly. Their voices cancel each other and you wouldn't be able to understand each of them. In highly loaded WCDMA cells, **cell breathing** effect arises in which all UE compete with each other and increase the power because of the increased uplink noise by many UEs. Now UEs at cell edge cannot win due to their distance and hence cannot be served by this cell coverage

The inner loop power control is based on target signal to interference ratio (SIR target) configured in the uplink

We have just touched concepts of UMTS planning. Now we will give same consideration to LTE.

LTE

LTE networks are deployed on B3 or DCS-1800 and B1 2100. Some operators with fewer sites (hotspots) and want large coverage may also opt for the digital dividend or DD-800 B20. However the DCS has larger bandwidth than the DD

The LTE planning takes the following steps

- Frequency planning
- PCI planning
- TAC planning
- PRACH planning

Before we go into details of those steps, let us take a look at LTE operations and start from the moment LTE enabled smartphone is powered on.

This process is shown below from powering on all the way to selecting a suitable cell. Please note that we are showing a simplified view of the general steps involved.

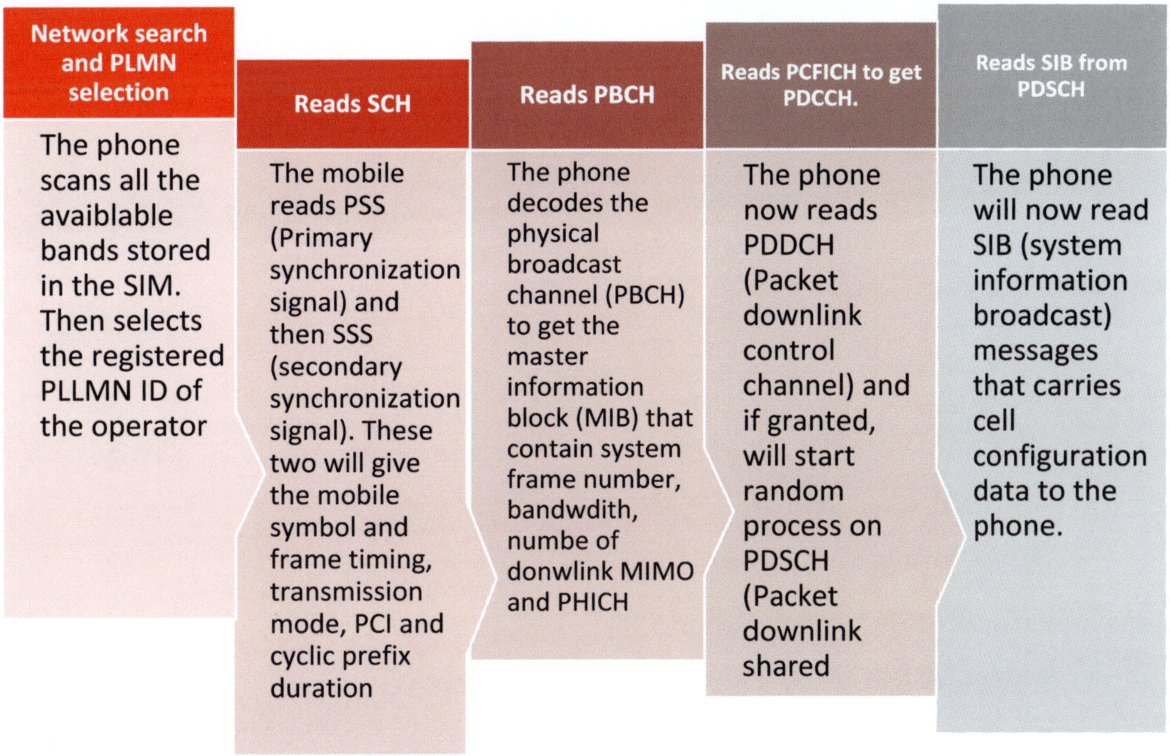

Network search and PLMN selection	Reads SCH	Reads PBCH	Reads PCFICH to get PDCCH.	Reads SIB from PDSCH
The phone scans all the avaiblable bands stored in the SIM. Then selects the registered PLLMN ID of the operator	The mobile reads PSS (Primary synchronization signal) and then SSS (secondary synchronization signal). These two will give the mobile symbol and frame timing, transmission mode, PCI and cyclic prefix duration	The phone decodes the physical broadcast channel (PBCH) to get the master information block (MIB) that contain system frame number, bandwdith, numbe of donwlink MIMO and PHICH	The phone now reads PDDCH (Packet downlink control channel) and if granted, will start random process on PDSCH (Packet downlink shared	The phone will now read SIB (system information broadcast) messages that carries cell configuration data to the phone.

Figure 9.14 Basic LTE cell selection process after phone power on

After these processes are completed and the phone receive downlink reference signal, the phone is said to have acquired an LTE suitable cell.

At this stage, the phone or more commonly called UE in LTE world does not have resources on uplink shared channel (PUSCH) to transmit data. The UE also does not have uplink control channel (PUCCH) to send uplink scheduling request for PUSCH resource. Thus the UE will go through a process called random access.

After random access completion, the phone will receive three parameters

- PUSCH resource element to transmit uplink data
- Timing advance (TA) to compensate for round trip delay
- An identifier to be used by the phone in the uplink subsequent RRC establishment process called C-RNTI (cell radio network temporary identifier)

The next step the UE will do is to establish RRC connection with the EUTRAN and then attach procedure to register with the MME and get an IP address to communicate. In this way UE context profile is created in the MME and default radio bearer established in the EUTRAN. After this the phone is said to be in RRC-connected and EMM-registered. These steps are illustrated below

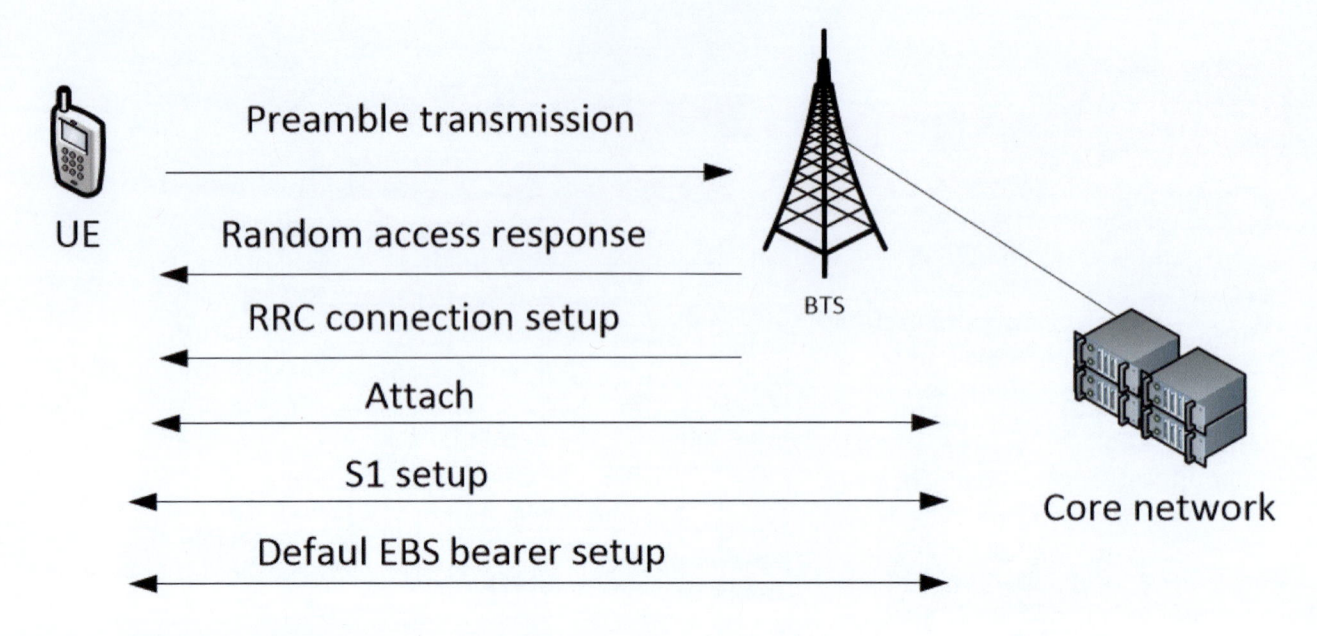

Figure 8.15 LTE process after phone power on until data transfer

Figure 9.16 summarizes steps taken by the UE from power until EMM-Registered, RRC connected.

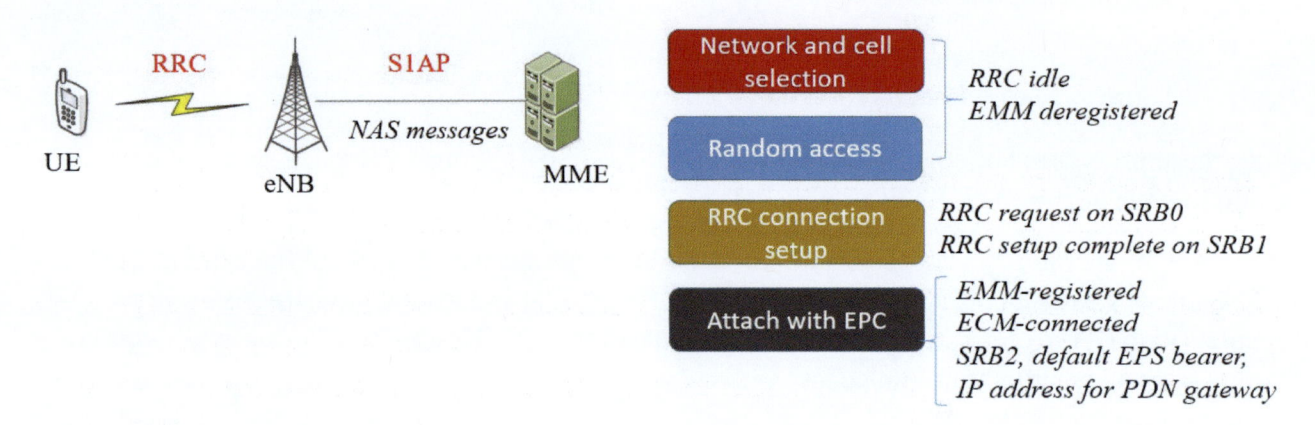

Figure 9 LTE power-on procedure steps

The UE has two EMM states:

Deregistered EMM state in which the UE location is unknown to the MME and thus UE cannot be paged.

Registered EMM state in which EPS default bearer is established and UE has IP address to make communication. EMM – registered has two further states. ECM-connected, RRC-connected in which UE location is known at cell level and UE performs cell change by handover. ECM-idle, RRC-idle in which UE location is known at tracking area (TA) level. UE transitions to connected state by (1) paging due to data arrival (2) triggering RRC connection request.

Different services required different quality of service (QOS) settings. For each APN there will be an associated default bearer. The PCRF dynamically generates QOS policies and sends them to the P-Gateway. The S-Gateway will covert S5 bearer from the P-Gateway to S1 bearer to the eNodeB. The HSS will then store subscribed QOS policies for each UE. The different LTE bearers is summarized in the following diagram

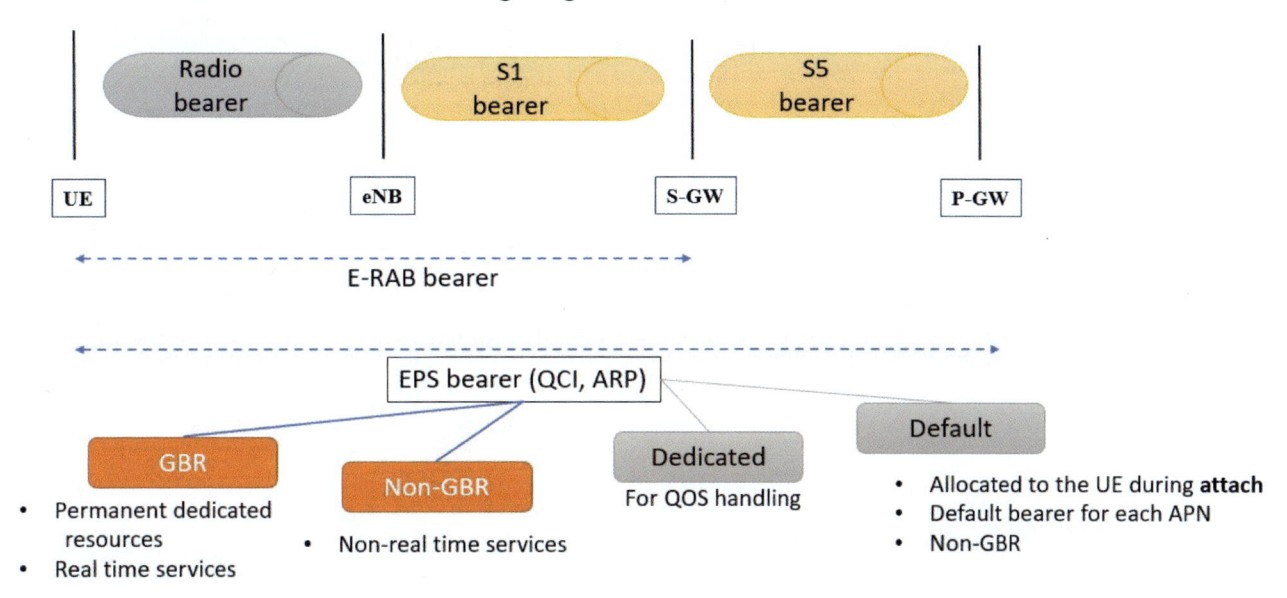

In Huawei implementation, the following parameters are set

QCI	1-9
ARPSchSwitch	ARPprioritySchSwitch-1
ARP1priority	GOLD
GoldUlSchPriorityFactor	1000

Since this book focuses on cellular air interface, it would be helpful if we look at what happens on the LTE radio interface. Data that passes on this air interface is called frame. This frame should conform to language understandable by both the eNodeB and the UE. This means the data frame should be formatted.

LTE radio frame is divided into many subcarriers. 12 subcarriers will form physical resource block (PRB) in the frequency domain. The subcarrier is also divided into 20 time slots in the time domain. 2 time slots will together form a sub-channel to be granted to the UE. This process is illustrated below

LTE radio frame

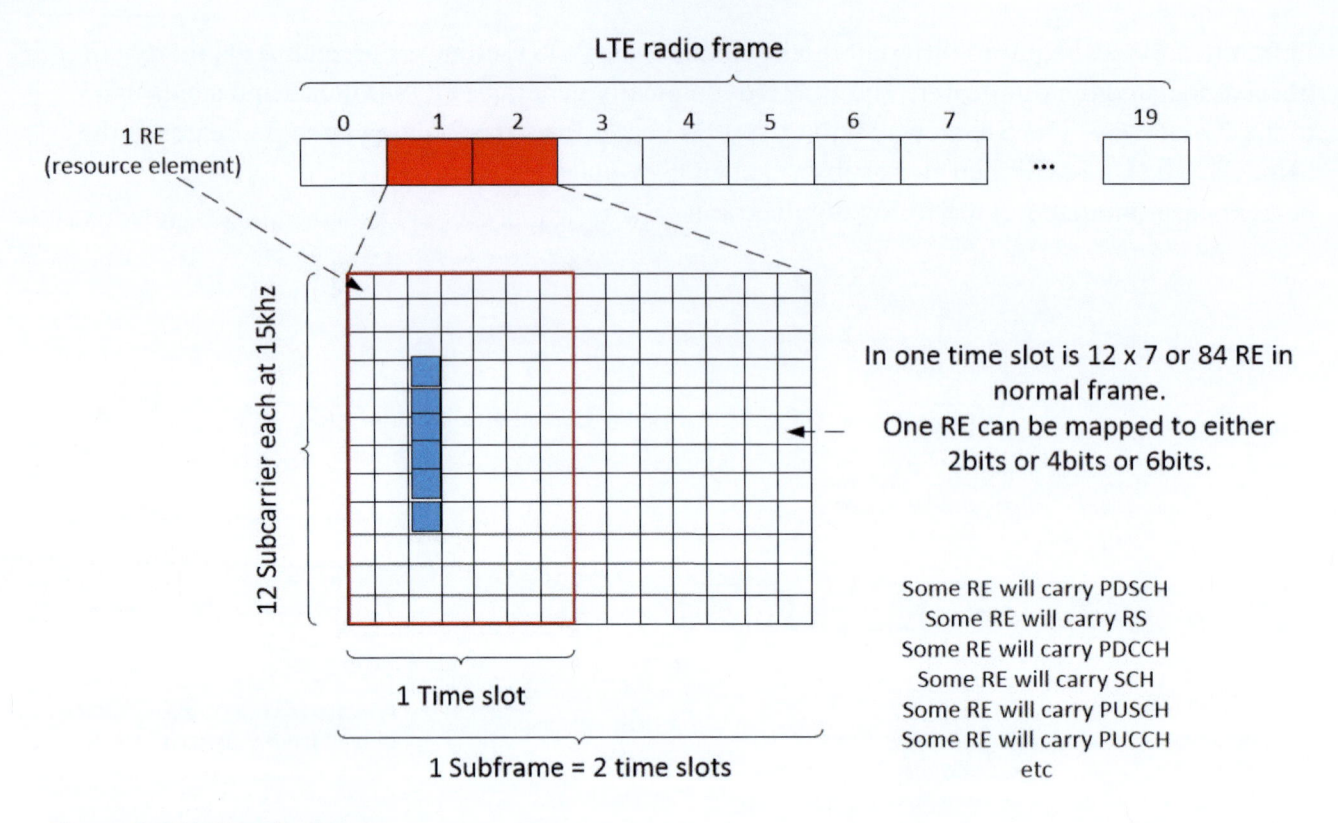

In one time slot is 12 x 7 or 84 RE in normal frame.
One RE can be mapped to either 2bits or 4bits or 6bits.

Some RE will carry PDSCH
Some RE will carry RS
Some RE will carry PDCCH
Some RE will carry SCH
Some RE will carry PUSCH
Some RE will carry PUCCH
etc

Figure 9.16 LTE radio frame

From this diagram we can infer the following

- One PRB is made up of 12 subcarriers and two time slots
- One time slot is made up of 7 symbols in normal frame or 6 symbols in extended frame
- The bandwidth of each subcarrier (RE) is 15 kHz. Hence bandwidth of PRB is 15 x 12 = 180 kHz
- The smallest unit of the LTE radio frame is resource element (RE). Each box in the grid above is resource element
- Not all those boxes or RE will carry customer data. Some will carry overhead control data
- In normal frame you will have total of 84 RE or symbols. The eNodeB scheduler will modulate each RE with either QPSK, 16-QAM, or 64-QAM in adaptive modulation
- QPSK will be assigned when the UE reports bad channel environment such as users in the cell edge where the signal is weak and interference is high. QPSK allows small throughput

LTE has flexible bandwidth configuration depending on capacity needs and more importantly spectrum availability. Each bandwidth will then have number of PRBs available.

As an example suppose that an RF planner allocates 20 MHz to a new eNodeB. Let us calculate theoretically expected throughput of this link.

In 3GPP specification release 8 (R8) LTE will have 64-QAM and 2 x 2 MIMO in the downlink. You will have 2 transmit antenna and 2 receive antenna in 2 x 2 multiple antenna system (MIMO)

In 64-QAM you will have (2^6) or 6 bits of data modulated by the RE

We said one time slot is made up of 7 symbols in the time domain and 12 subcarriers in the frequency domain. Hence 1 time slot = 7 x 12 RE

In LTE two time slots constitute one sub-channel or sub-frame. Hence one sub-frame contains 2 x 7 x 12 = 168 RE

In LTE 20 MHz deployment, we have 100 PRB. Hence total RE = 168 x 100 = 16800

Since each RE is modulated by 6 bits in 64-QAM, 100 RE will be modulated by 16800 x 6 = 100,800 bits. The question is how many bits per second. Remember LTE radio frame is 20ms

Hence in 20 MHz radio frame theoretical throughput for one antenna transmitter is

100,800 bits per millisecond

1 s = 1000 ms, hence 100,800,000 bits per second is sent every millisecond radio frame.

Converting into megabits per second, we have 100.8Mb/s for single transmitter antenna on the eNodeB

One thing to remember is that not all the 100.8Mb is user plane data. Some will be overhead control plane data depending data on how much features and licenses are configured on the eNodeB. The configured PRB is broadcasted to the phone on master information block (MIB) block

There are different SIB messages in LTE and each one delivers specific information to the phone. The following table summarizes some of those SIBs the RF planner is required to understand their function and their contents

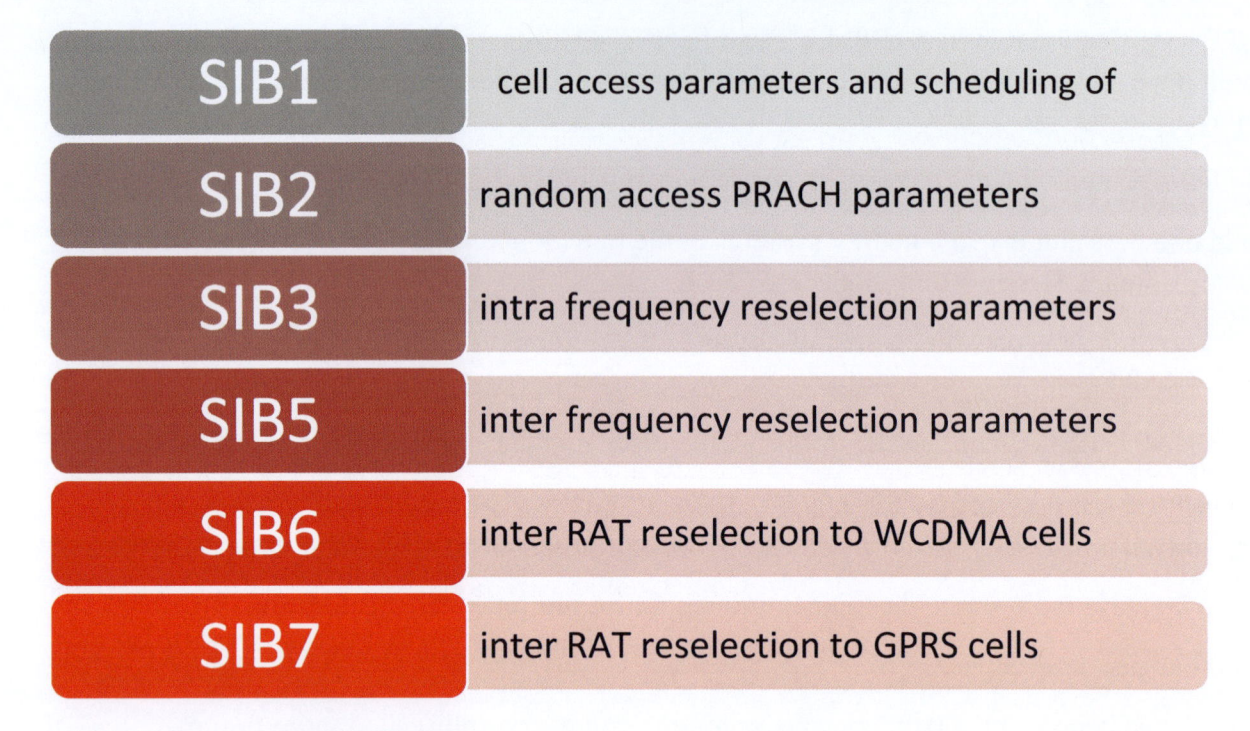

SIB1	cell access parameters and scheduling of	
SIB2	random access PRACH parameters	
SIB3	intra frequency reselection parameters	
SIB5	inter frequency reselection parameters	
SIB6	inter RAT reselection to WCDMA cells	
SIB7	inter RAT reselection to GPRS cells	

Figure 9.17 some LTE SIB messages

Correct configuration of those SIB messages in the eNodeB is simplified by understanding the contents of these 3GPP messages. Overview of main parameters contained in the SIB messages in which the RF planner is required to plan, tune, and configure is shown below

Table 9.9 LTE SIB contents for cell selection and reselection

SIB-1	SIB-2	SIB-3
PLMN-ID:	numOfRApreambles:	sIntraSearch:
TAC: 3001	PowerRampingStep:	sNonIntraSearch:
Cell Not barred	PreambleIntitialReceivePower:	ThreshServLow
QrxLevMin:	PreambleTransMax:	q_Hyst:
	MAC ContentionResolutionTimer:	CellReselPriority:
	MaxHARQ_Msg:	P_max:
		t_Reselection EUTRA:

SIB-5	SIB-6	SIB-7
DL-EARFCN:	UTRA carrier:	Start ARFCN:
QrxLevMin:	CellReselPriority:	BandIndicator:

P_MAX:	ThreshXlow:	ARFCN List:
ThreshXhigh:	QrxLevMin:	CellReselPriority:
ThreshXlow:	t_Reselection UTRA:	QrxLevMin:
CellReselPriority:	P_MAX:	ThreshXlow
MeasBW:	QqualMin:	
t_Reselection EUTRA		

The UE will receive the SIB messages broadcasted in the cell on physical downlink shared channel (PDSCH) as illustrated below

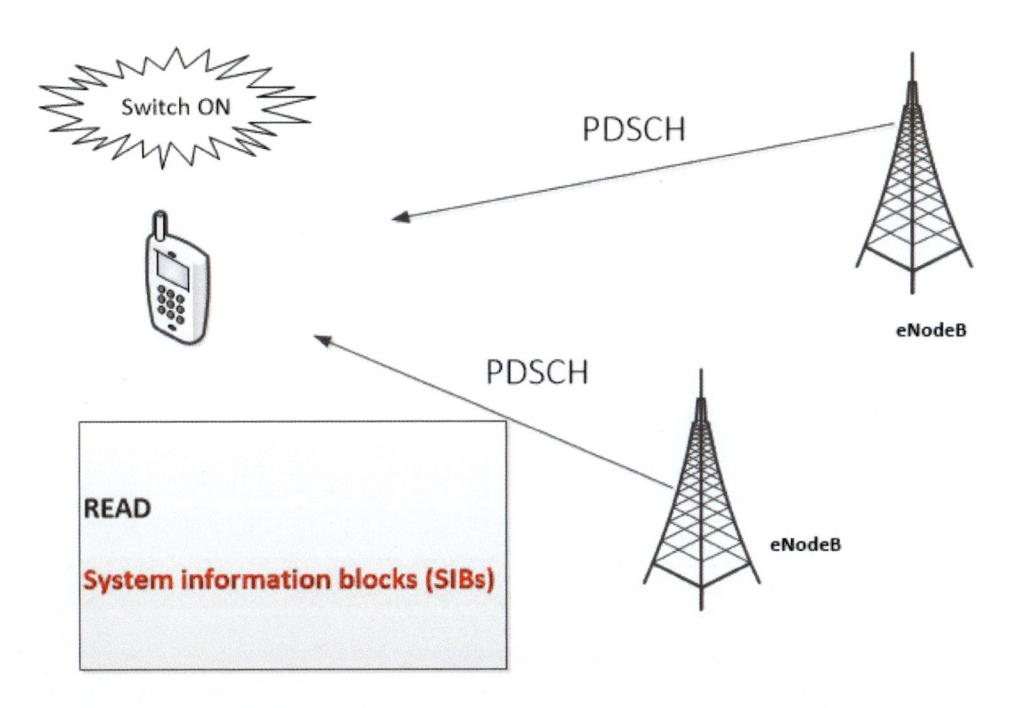

Figure 9.18 LTE SIB broadcast to the phone

LTE uses single frequency deployment on all cells. Hence frequency reuse is 1. In congested areas where cell edge interference is high (manifested by small SINR), two frequency blocks may be licensed from different LTE bands supported in the region.

When deploying LTE, spectrum is very scarce as the required bandwidth is very large. For example a typical LTE single frequency deployment will required 20 MHz channel. If the operator licensed the 1800 band which has only 75Mhz bandwidth. If the first 15 MHz is taken away by GSM-1800, then only 60 MHz will remain for three operators to divide among themselves.

This would be challenge if new operators decides to enter the market or incumbent operators decide to upgrade from 20 MHz to 40 MHz carrier aggregation in LTE-Advanced. The other problem is that B1 is already occupied by 3G system which required 10 MHz capacity in DC-HSPA+ systems that support 42Mbps.

One possible solution for the new operators is to license 20 MHz from LTE-2600 band. This will come with reduced coverage. If the operators had decided to deploy micro-cells instead of macro-cells, deploying many cells will be optimal for the 2600 band.

Table 9.10 B7 for micro cell high capacity LTE

| LTE-2600 | 2500-2570 | 2620-2690 | 70 | B7 |

The PCI planning is one of the most important parameters planned for the LTE cell. It serves the same purpose as the UMTS PSC codes. It differentiates downlink transmission from different cells. In LTE total of 504 PCI are available and reuse distance should be reasonable in case they are run out in highly dense LTE micro and femto cells.

An example of PCI planned and assigned to cells is shown below

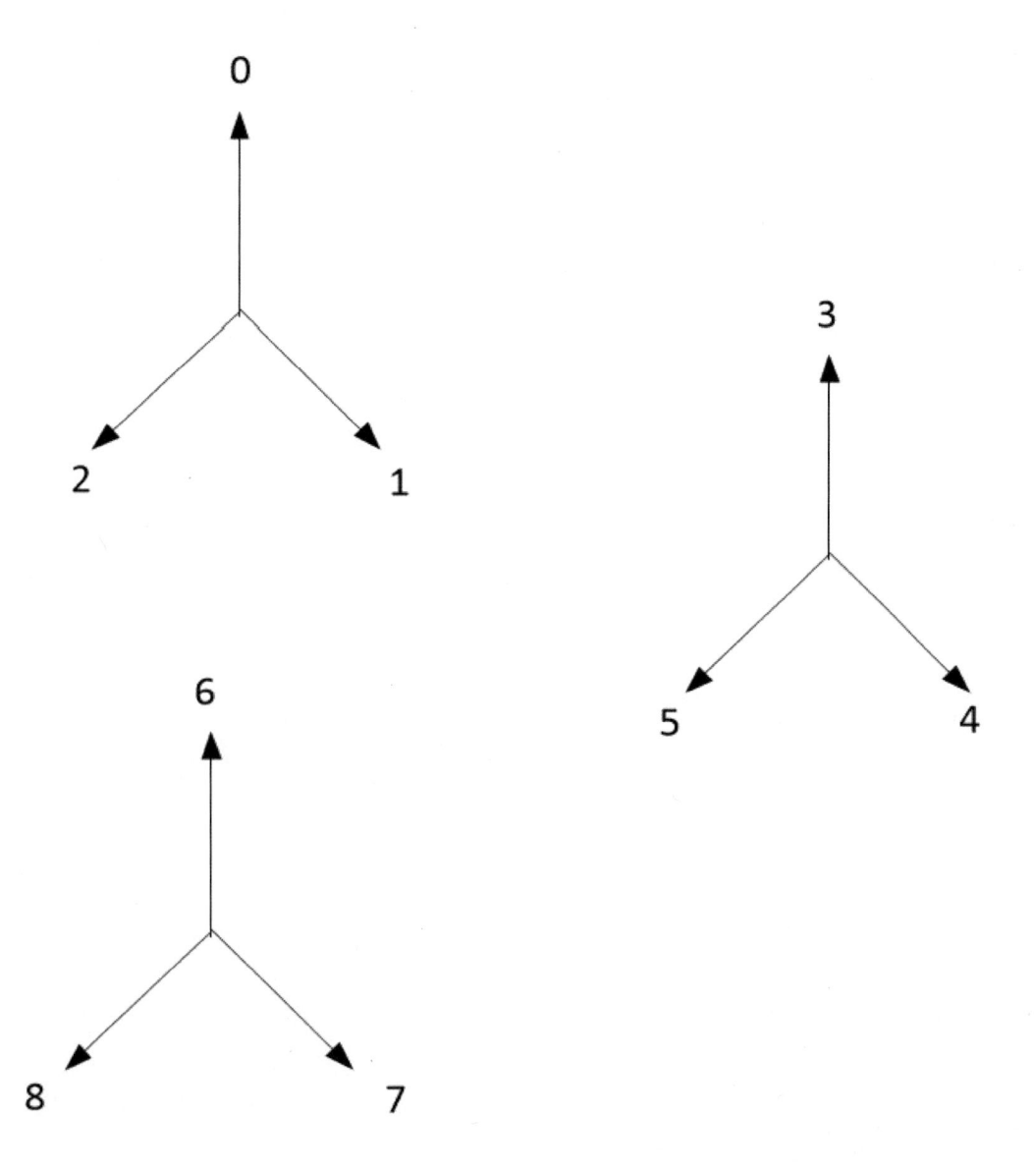

Figure 9.19 Example of LTE PCI plan

The LTE phone will decode PCI after reading PSS and SSS signals in the initial network latch process. This means PCI is formed as

$PCI = 3 \times SSS + PSS$

Where PSS can take the range 0, 1, 2 and SS can take the range 0 to 167

Hence the maximum PCI value is 3 (167) + 2 = 503

If PCI planning is done wrong either manually or in the planning tool, serious performance issues will bring down the whole LTE network. The PCI planning can go wrong in two ways. PCI collision and PCI confusion as explained in the figure below

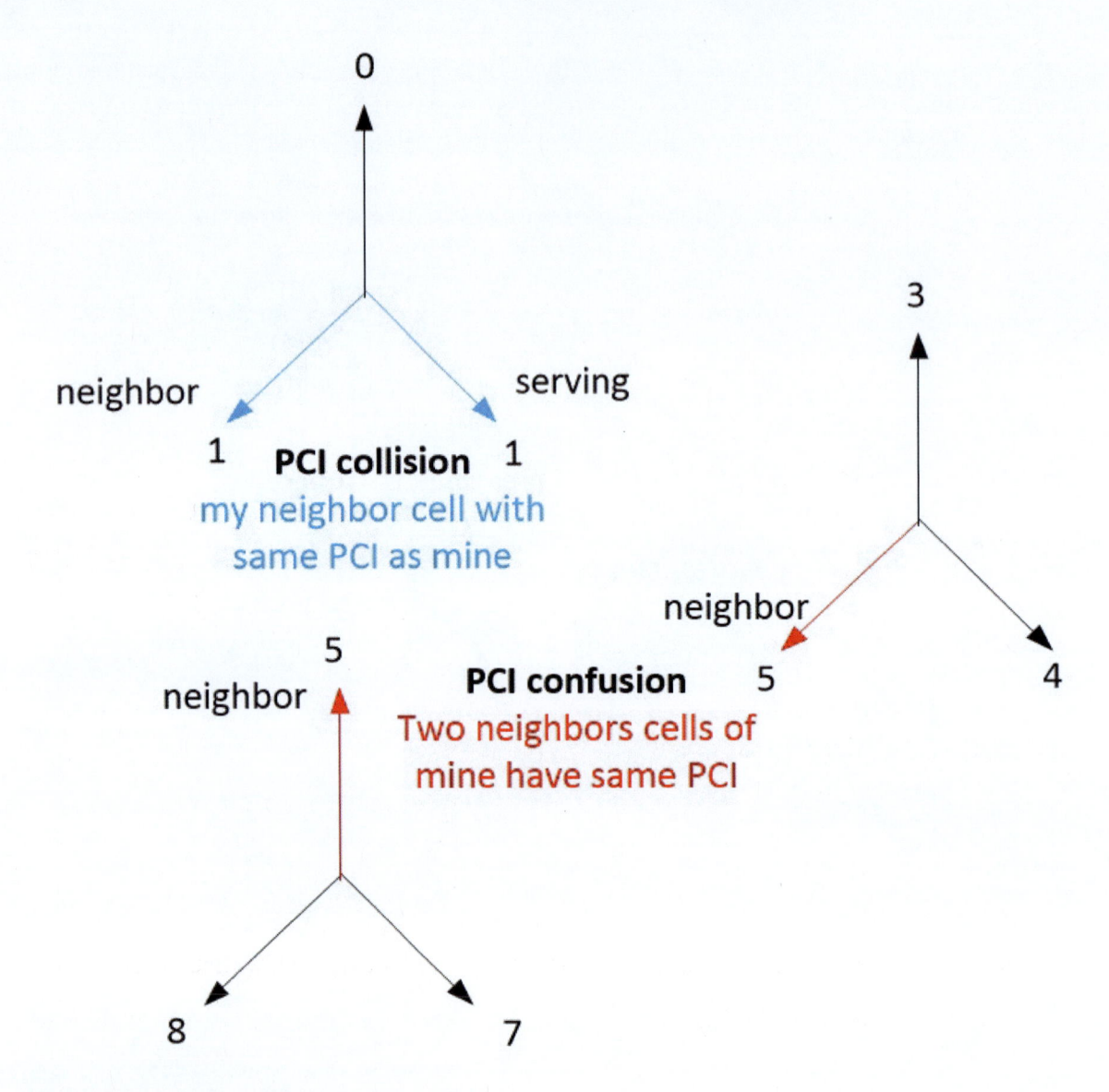

Figure 9.20 LTE PCI collision vs PCI confusion

Another important planned for the LTE is tracking area code (TAC). A group of cells are planned for a given tracking area and each tracking area (TA) is assigned with tracking area code (TAC)

The idea of TA is similar to GSM LAC and WCDMA RAC planning. The network is divided into smaller areas to reduce paging load. Paging is signaling messages sent to all cells in a TAC to wake the phone located in that TA to received data. To save phone battery power there is standby time called DRX cycle (discontinuous reception) in which the UE sleeps. Then after every DRX cycle the phone wakes up to listen to the paging channel.

Every time the phone moves between different TA, it updates the MME its current location by sending TAU (tracking area update) message. This message goes directly to the MME via the EUTRAN. It is therefore called NAS (Non access stratum) signaling as opposed to AS (access stratum) such as RRC (Radio Resource Control) messages which terminate at the EUTRAN.

An example of LTE network with TAC planning is shown below

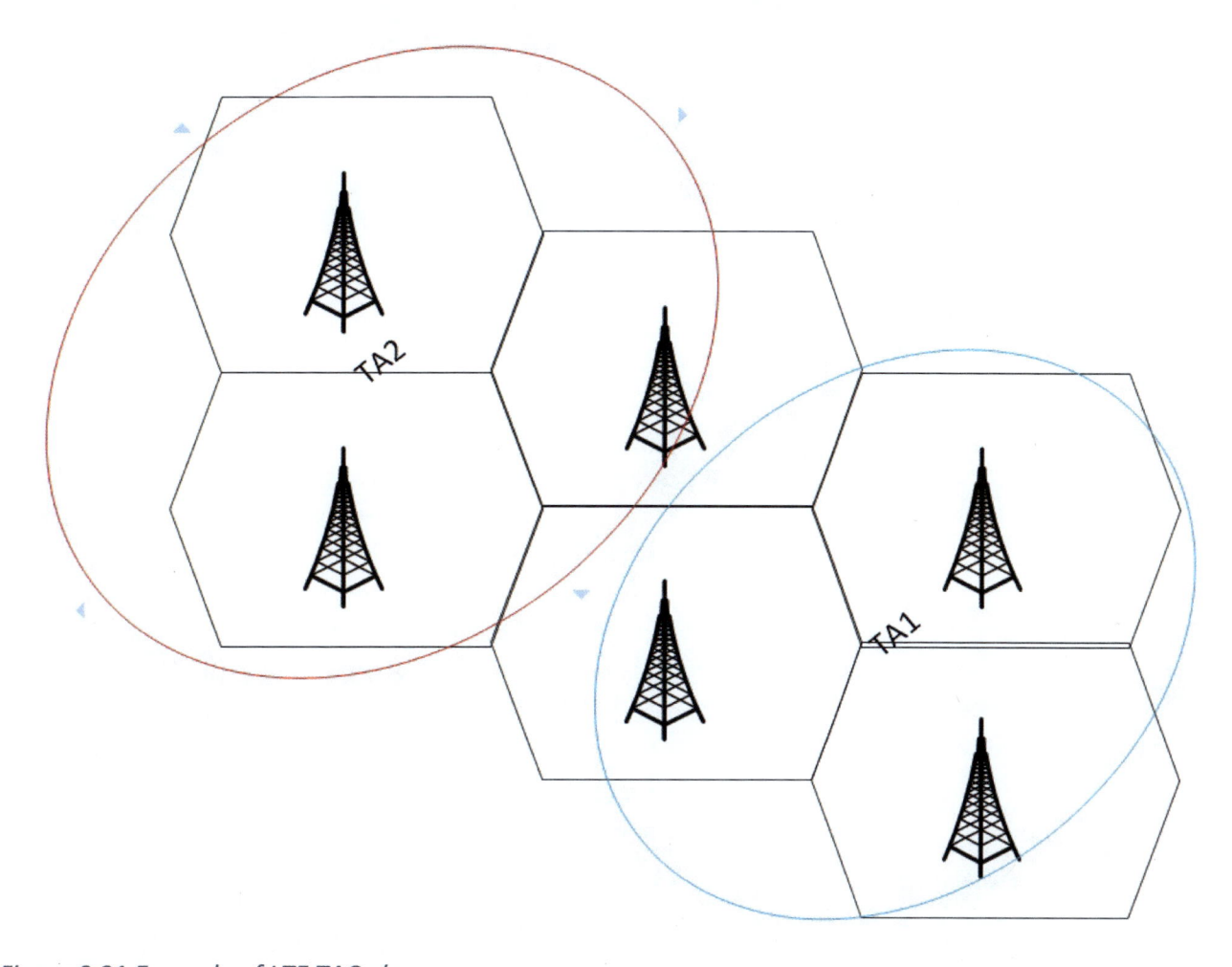

Figure 9.21 Example of LTE TAC plan

In the configuration pane of the eNodeB, all cells within one TA are given one TAC. For example we could give TA1 area cells with TAC code of 1000 and cells in TA2 given TAC of 1010

Later we will see issues seen in newly deployed LTE network of TAU reject messages and ways to fix it.

The final part of LTE PRACH planning is also important. We will touch basic things telecom managers and planners need to know.

The basic idea behind random access process or PRACH (physical random access channel) is to achieve uplink synchronization between the phone and the UE so that uplink resources such as RRC would be granted to the phone. Whenever the phone loses uplink synchronization for example when turned off and then on, it sends special message to the eNodeB called random access preamble. The random process usually follows four steps as shown in figure 9.22

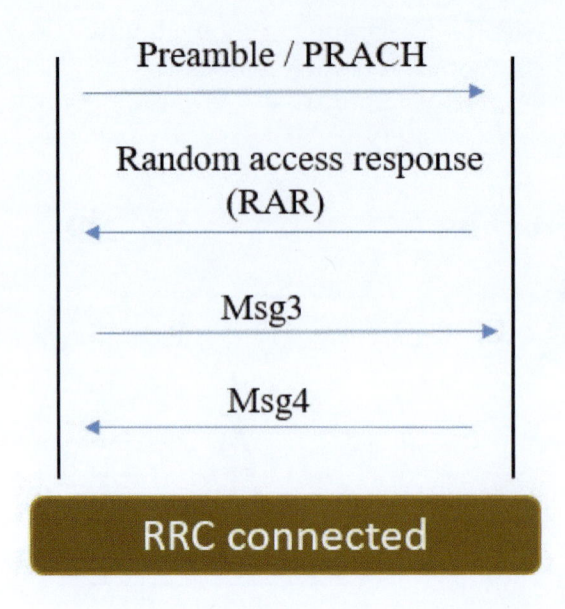

Figure 9.22 LTE random access procedure

There are 64 such preambles available in the cell for the phone to pick randomly. When multiple phones pick same preamble randomly, the collision that results is called contention-based PRACH. The process that the eNodeB uses to solve when it receives same preamble from different phones is called contention resolution.

There are special cases such as during handover procedure in which the eNodeB assigns a reserved PRACH preamble to the phone so that it does not necessarily need to pick one by itself. Collision is minimized in this case and it is called contention-free random access

Now the UE has received master information block (MIB) that tells which system frame number in the next transmission. The question is how will the phone know the RE to use for PRACH?

Location of PRACH RE is indicated in two parameters in SIB2
- *PRACH configuration index* - specifies time-domain of the RE
- *PRACH frequency offset* - specifies frequency-domain location of the RE

Each LTE cell supports 64 preamble sequences .The UE generates these sequences using different combinations instructed by *root sequence index* parameter in SIB2. It is the job of the RF planner to plan those PRACH parameter and configure them on the LTE cells

The question is how many root sequences are needed per cell for a certain coverage? It is decided by another parameter called cyclic shift or Ncs

Table 9.11 LTE PRACH root sequence index vs cell size

zeroCorrelationZoneConfig N_{CS}	No. of cyclic shifts	PRACH preamble sequences per root sequence	Root sequences required per cell	Cell range [km]
1	13	64	1	0.76
2	15	55	2	1.04
3	18	46	2	1.47
4	22	38	2	2.04
5	26	32	2	2.62
6	32	26	3	3.47
7	38	22	3	4.33
8	46	18	4	5.48
9	59	14	5	7.34
10	76	11	6	9.77
11	93	9	8	12.2
12	119	7	10	15.92
13	167	5	13	22.78
14	279	3	22	38.8
15	419	2	32	58.83
0	839	1	64	118.9

As an example if the *cyclic shift* number if 4, number of preambles per root sequence = 839/Ncs = 839/22 = 38 which would give approximately 2km cell coverage. The table also says two roots sequences needed to be configured in each cell.

The resultant *rootsequenceindex* plan for typical cell configuration is shown below

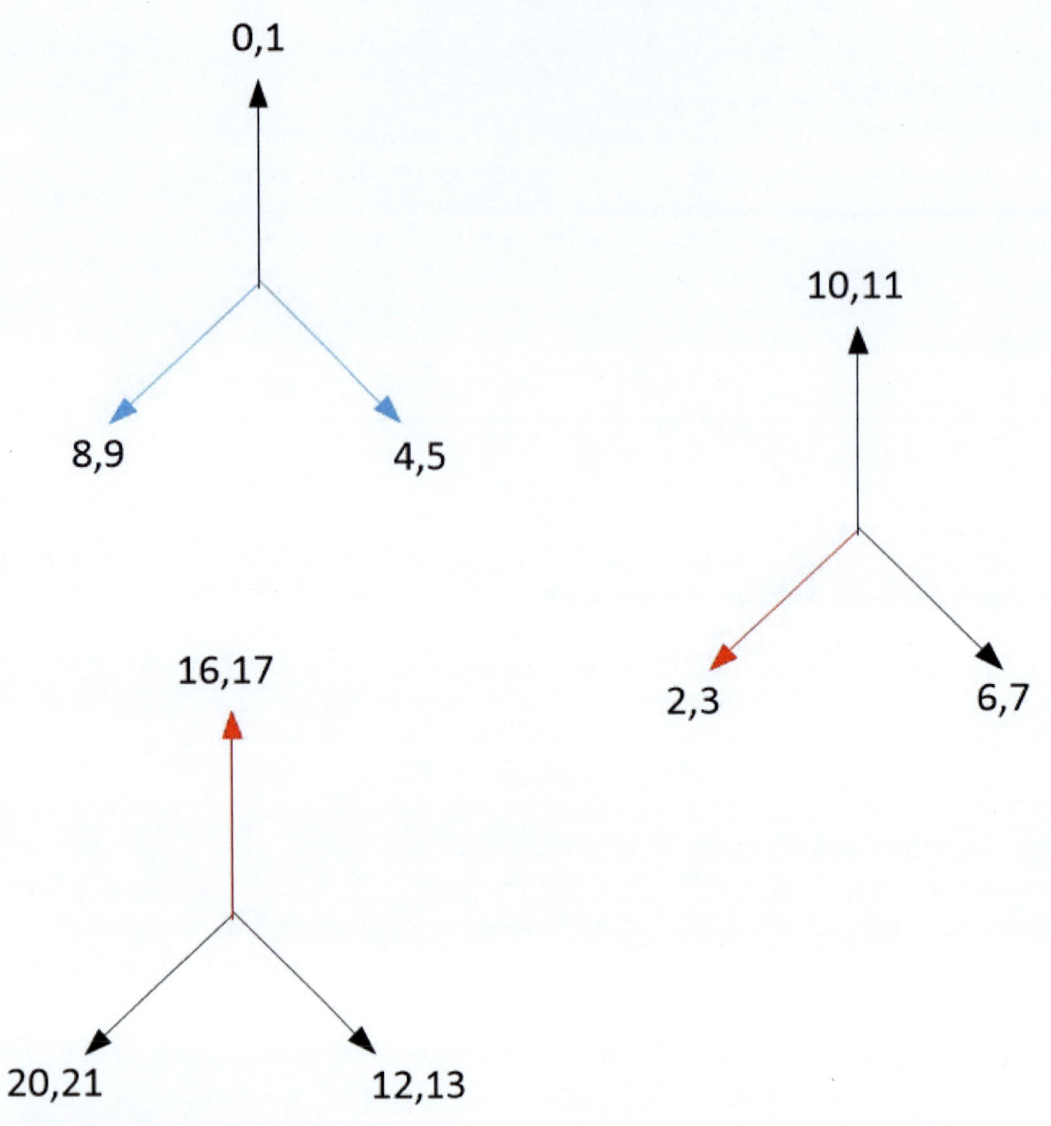

Figure 9.23 Example of LTE PRACH plan

One of the main factors which effect LTE cell coverage is preamble format used. There are four preamble formats in LTE and which format to be used is determined by the parameter *configurationIndex* in SIB2. More information is found in the 3GPP standard TS.36.211 table 5.7.1-
A typical PRACH parameters in SIB2 for certain operator is summarized below

Table 9.12 Typical LTE cell PRACH parameters

SIB2 parameter	Note	Example settings
Number of random access preambles	To be generated by the phone randomly using cell instructions	64
Power ramping step	Increment increase of preamble power from initial received target	1dB

	power	Actual is 2dB
Preamble initial received target power	Starting preamble transmit power	-106dBm
Preamble re-transmission maximum	How many times preamble to be transmitted if response is not heard from the cell	n8
Random access response window size	Specifies subframe interval between RACH request and RACH response by the network	SF32
MAC-contention resolution timer	Timer for contention resolution to complete. If not new preamble is sent again.	SF8
PRACH configuration index	Determines preamble format for the cell	0 - 63
PRACH frequency offset	Frequency location of the preamble RE	0 – 94
Root sequence index	Broadcasted to the cells to tell the phone how to generate the preambles	0 - 837

From the above table when we say subframe we mean 1ms TTI (transmission time interval)

Preamble power transmission uses open loop power control algorithm as opposed to closed loop power control used in PUSCH/PUCCH. Open loop power control does not need feedback from the cell.

For the phone to estimate what power to transmit the preamble, it uses the following information

- Measured signal level of the downlink reference signal or RSRP
- Preamble initial received power expected by the cell extracted from SIB2
- Calculation of pathloss between the phone antenna and cell antenna

Then the PRACH transmit power is calculated from the following formula

$PRACH\ power = min(P_{Max}, PreambleReceiveTargetPower + pathloss)$

As an example assume the downlink reference signal (RS) transmitted from cell antenna is 18dBm and the initial receive target power is set to -106dBm. Assume also maximum phone TX power allowed is 24dBm and RSRP at the phone location is -100dBm

Pathloss = transmit power − receive power = RS power − RSRP = 18 − (-100) = 118

PRACH power = min (24, − 106 + 118) = min (24,12)

This indicates the phone should send the preamble at transmit power of 12dBm so the cell receives it at -106dBm successfully. This example also shows that as RSRP decreases (for example in weak coverage scenario), the preamble transmit power also increases until power is exhausted and random access fails. Hence one of the solutions of bad network accessibility KPI is to improve the coverage.

Once random access response from the cell is successful the phone is now synchronized in the uplink and can proceed to requesting RRC connection link on the PUSCH

But before that let us talk about a little bit about LTE closed loop power control planning in PUSCH/PUCCH channels. PUSCH (physical uplink shared channel) carries customer data and PUCCH (packet uplink control channel) carries signaling messages.

In SIB2 certain parameters should be set for closed loop power control of these uplink channels. As an example the RF planner is required to set **PoNominalPUSCH** and **PoNominalPUSCH**.

We will touch these parameters a little bit in the cell coverage optimization section later in table 9.41

In LTE different transmission modes are supported based on **MIMO** settings. Multiple input multiple output (MIMO) transmission is used to combat channel quality degradation and increase user throughput. In R8, downlink 2 x 2 MIMO deployment is more common in the initial rollout

TM2	Open-loop transmit diversity
TM3	Open-loop spatial multiplexing

If adaptive MIMO switching is configured, *spatial multiplexing* is used at cell center
To boost throughput (different data on same resource block). *Transmit diversity* is used at cell edge to Increase SINR and coverage (same data on same resource blocks)

Spatial multiplexing at cell center

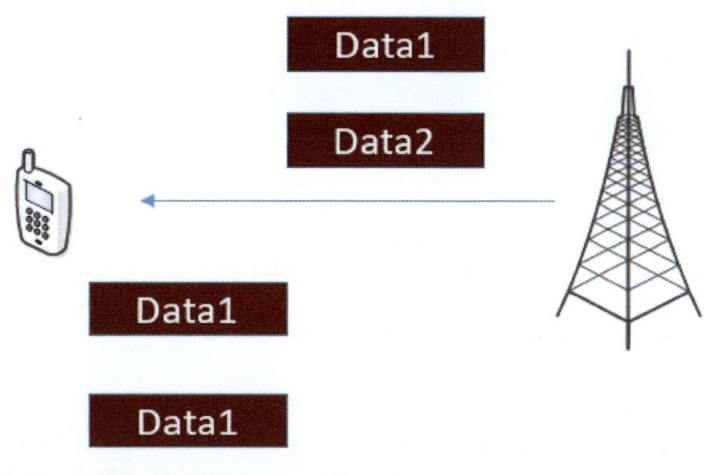

Transmit diversity at cell edge

Figure 9.24 TM2 and TM2 MIMO modes

The eNodeB can adaptively configure different transmission modes for different UEs in a cell and the UE will receive MIMO settings in RRC_connection_setup

TxRxMode	2T2R
InitialMimoType	ADAPTIVE
CrsPortNum	CRS_PORT_2
CQIPERIODADAPTIVE	ON
MimoAdaptiveSwitch	OL_ADAPTIVE
SrsCfgInd	TRUE

Interoperability

Operators who invest 2G/3G/4G technologies need to plan seamless co-existence and mobility management among the technologies. That means a user in LTE cell should be able to be served by that cell. If the user moves out of LTE cell, it should seamlessly get neighboring HSPA cell. If both LTE and HSPA coverage is moved out, the user should still be serviced on the wide coverage GSM cell.

In this section we shall talk about the following mobility RF planning

- GSM – WCDMA interoperability
- GSM – LTE interoperability
- WCDMA – LTE interoperability

To plan mobility management parameter, the state of the user should be taken into account

- In idle mode, the user or phone will do cell selection/reselection. Hence we plan cell reselection parameters which will be broadcast to the phone by the base station in SIB messages. In idle mode the phone will do location update
- In connected mode, the phone will send measurement report to do handover. In WCDMA/LTE the phone reports events when threshold parameters in the event configuration are exceeded.

During cell reselection the phone will measure the receive level of the serving cell. If the receive strength and quality of the serving cell exceeds configured thresholds, then phone will reselect another suitable cell (provided neighbor relationship between the serving and target cells exists)

An example of signal threshold for a typical macro cell used in cell reselection is summarized below. Note that different cell sizes requires different reselection thresholds. So the cells will be planned differently depending on cell coverage radius.

Table 9.13 Typical cellular operator signal thresholds in cell reselection

		target cell		
		GERAN	UTRAN	EUTRAN
Serving cell	**GERAN**	$RSSI \leq -95$	$RSSI \leq -95$ AND $RSCP \geq -100$	$RSSI \leq -95$ AND $RSRP \geq -110$
	UTRAN	$RSSI \geq -95$ AND $RSCP \leq -100$	$EcNo \leq -12dB$ AND $RSCP \leq -100$	$RSCP \leq -95$ AND $RSRP \geq -100$

	EUTRAN	$RSSI \geq -95$ AND $RSRP \leq -110$	$RSCP \geq -100$ AND $RSRP \leq -110$	$RSRP \leq -110dB$ AND $RSRQ \leq -18dB$

From the table above, we mentioned signal level and quality used in each technology. The following table summarizes their meaning and ranges

Table 9.14 GSM UMTS LTE signal terminology and range

Parameter name	Meaning	Range	Technology
RSSI	Receive signal strength indicator	-55 to -104dBm	GERAN
RSCP	Receive signal code power	-65 to -115dBm	UTRAN
EcNo	Code energy over noise density	-3 to -15dB	UTRAN
RSRP	Receive signal reference power	-70 to -120dBm	EUTRAN
RSRQ	Receive signal reference quality	-3 to -19.5dB	EUTRAN

Signal level and quality are measured by all phones in the cell. Each phone continuously or periodically reports these levels in the uplink. For example in LTE the UE will send CQI (channel quality indicator) report to the eNodeB containing reference signal SINR. The eNodeB scheduler will then allocate adaptive modulation and hence corresponding throughput according to channel conditions which is random. That is why users in LTE cell edge have lower QPSK and throughput than users close to the eNodeB who enjoy much higher throughput of 64-QAM because of good CQI report.

Intra RAT cell reselection

Intra RAT or same radio access technologies maybe studied as

- GERAN - GERAN reselection
- UTRAN – UTRAN reselection
- EUTRAN – EUTRAN reselection

To discuss EUTRAN to EUTRAN cell reselection, we select a case in which the operator licensed to frequency blocks from B3 and B20. This decision was made to reduce cell edge interference and their network arrangement is as follows

From network planning tool all three sites were found to be neighbors. Since the sites only A and C use same band, the remaining neighbors (A-B, B-C) use different bands.

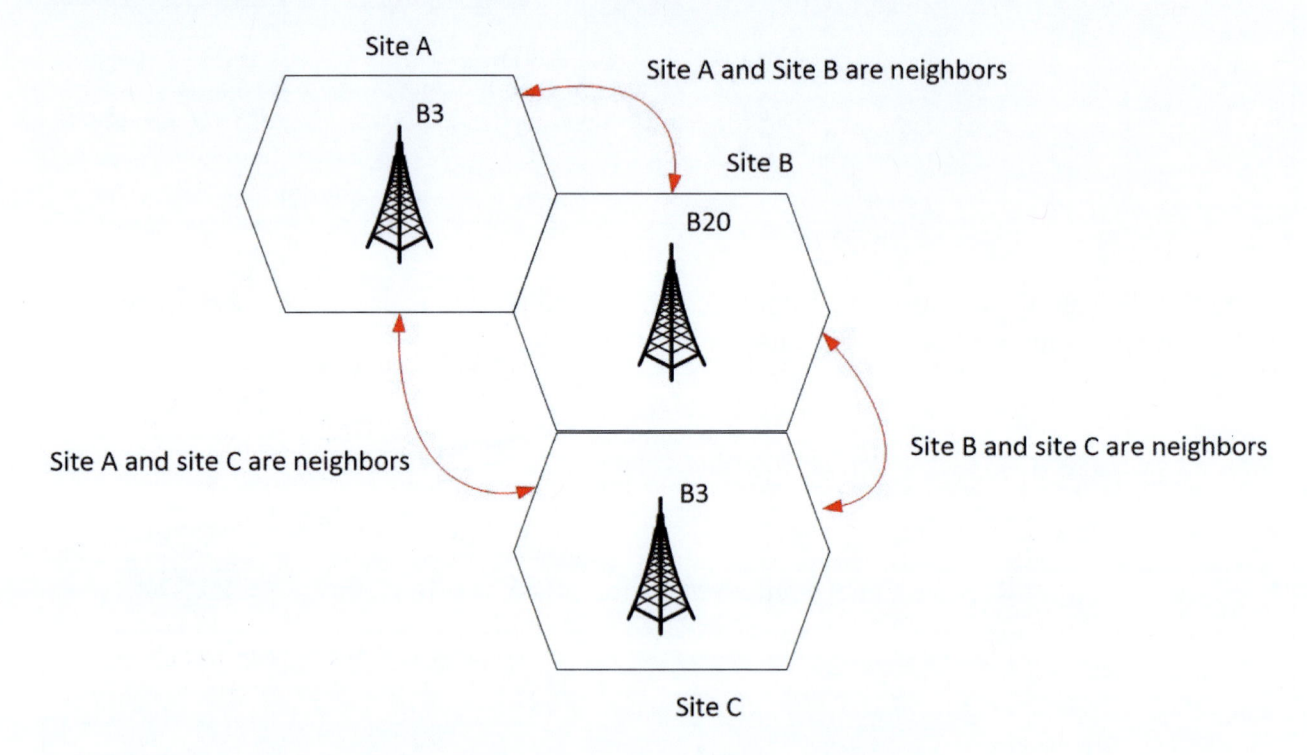

Figure 9.25 Example of LTE intra frequency vs inter frequency cell reselection

From this scenario we therefore need to configure reselection parameters as

- Intra frequency cell reselection parameters between sites A and C which are broadcasted in LTE SIB3
- Inter frequency cell reselection parameters between sites A and B, and sites B and C which are broadcasted in LTE SIB5

To make this design easy, LTE cells with same frequency are given same priority while those that are on different carrier are given different priority

In cellular system priority level from 0 to 7 is used with 0 being the lowest priority RAT (radio access technology) and 7 being the highest priority RAT. For example a certain operator with different option in mind might use the following priority levels

Table 9.15 RAT priority

Priority	RAT
1	GSM 900
2	GSM 1800
4	WCDMA

5	HSPA
6	LTE B20
7	LTE B3

With this table in mind, now our LTE B20 will have lower priority than LTE B3. This mean cells A and C will have higher priority than cell B. This in turn means as follows

Table 9.16 LTE inter frequency reselection priority

Reselection from high priority to low priority inter frequency	**From cell A to cell B**
Reselection from high priority to low priority inter frequency	From cell C to cell B
Reselection from low priority to high priority inter frequency	From cell B to cell A
Reselection from low priority to high priority inter frequency	From cell B to cell C
Reselection from equal priority intra frequency	From cell A to cell C and vice versa

The idea behind cell reselection based on RAT priority is that a phone in continuous coverage of LTE/WCDMA/GSM will always camp on the LTE cell as it provides the best user experience. If LTE is not available for the next cell reselection, it chooses the next highest priority which is always WCDM. If the operator has deployed many LTE cells either intra frequency or inter frequency, it is recommended to adjust and tune cell reselection parameters so that the user always stays on LTE. This RAT priority concept is further shown below

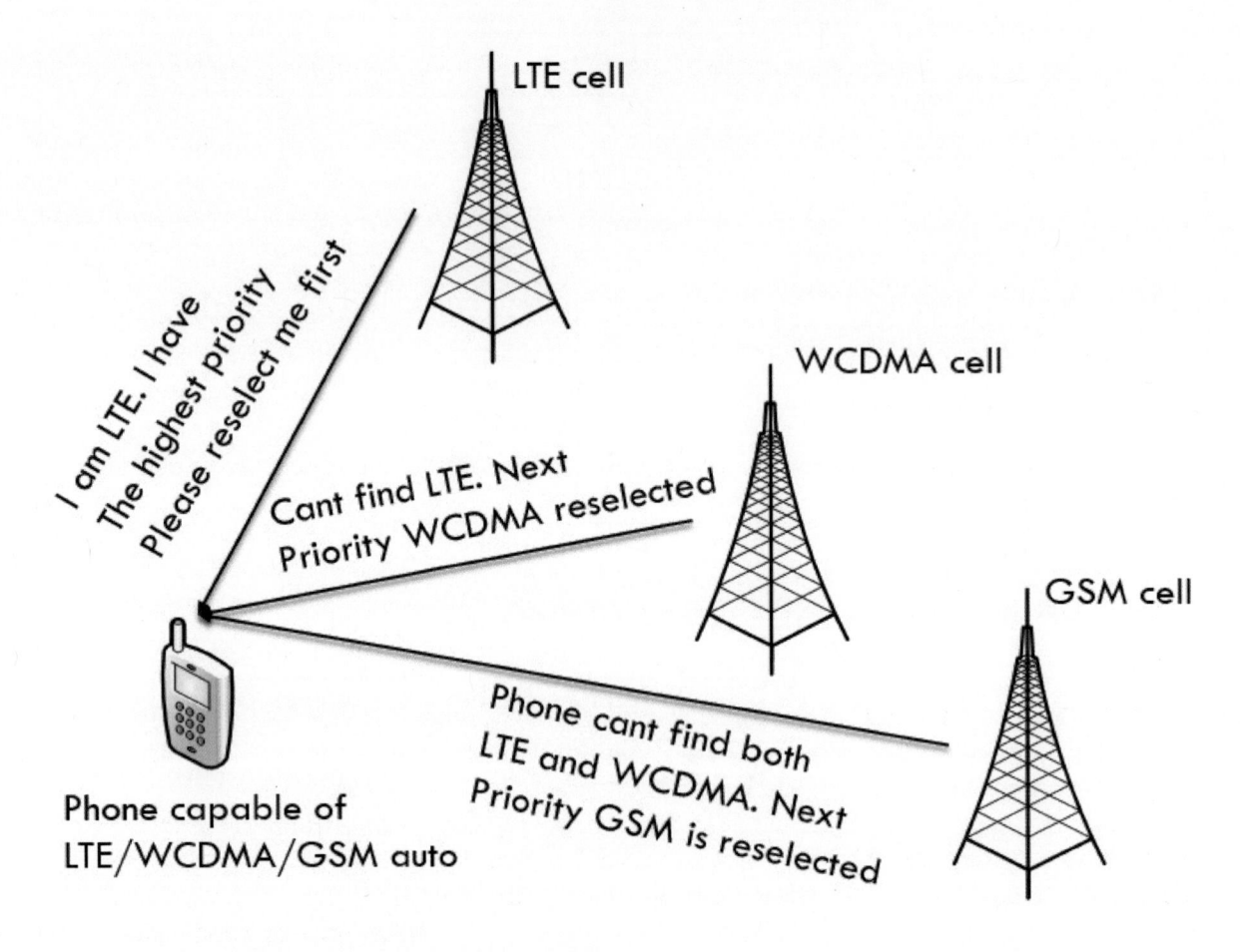

Figure 9.26 cell reselection based on RAT priority

In LTE cell reselection configuration, first the minimum signal level the phone can access the LTE cell is specified in SIB1. This is called qRxLevMin and can take up to -120dBm for large macro cells or -110dBm for small cells. The phone will select the LTE cell only if the measured RSRP by the phone is greater than this threshold. This threshold presents a convenient tool to the RF planner for adjusting the cell coverage.

Then as the phone or user is moving around the city either walking at low speed or on high speed driving, the mobility of the user must be respected. LTE cells in which neighboring relationship is established must be reselected from one to next.

To save phone battery power, it will not always do neighbor measurement. It will start neighbor intra RAT measurement when a certain threshold called sIntrasearch is exceeded. It will also start to measure other inter frequency LTE cells or inter RAT cells when certain threshold called sNonIntrasearch is exceeded.

The following table clarifies the above lines of text

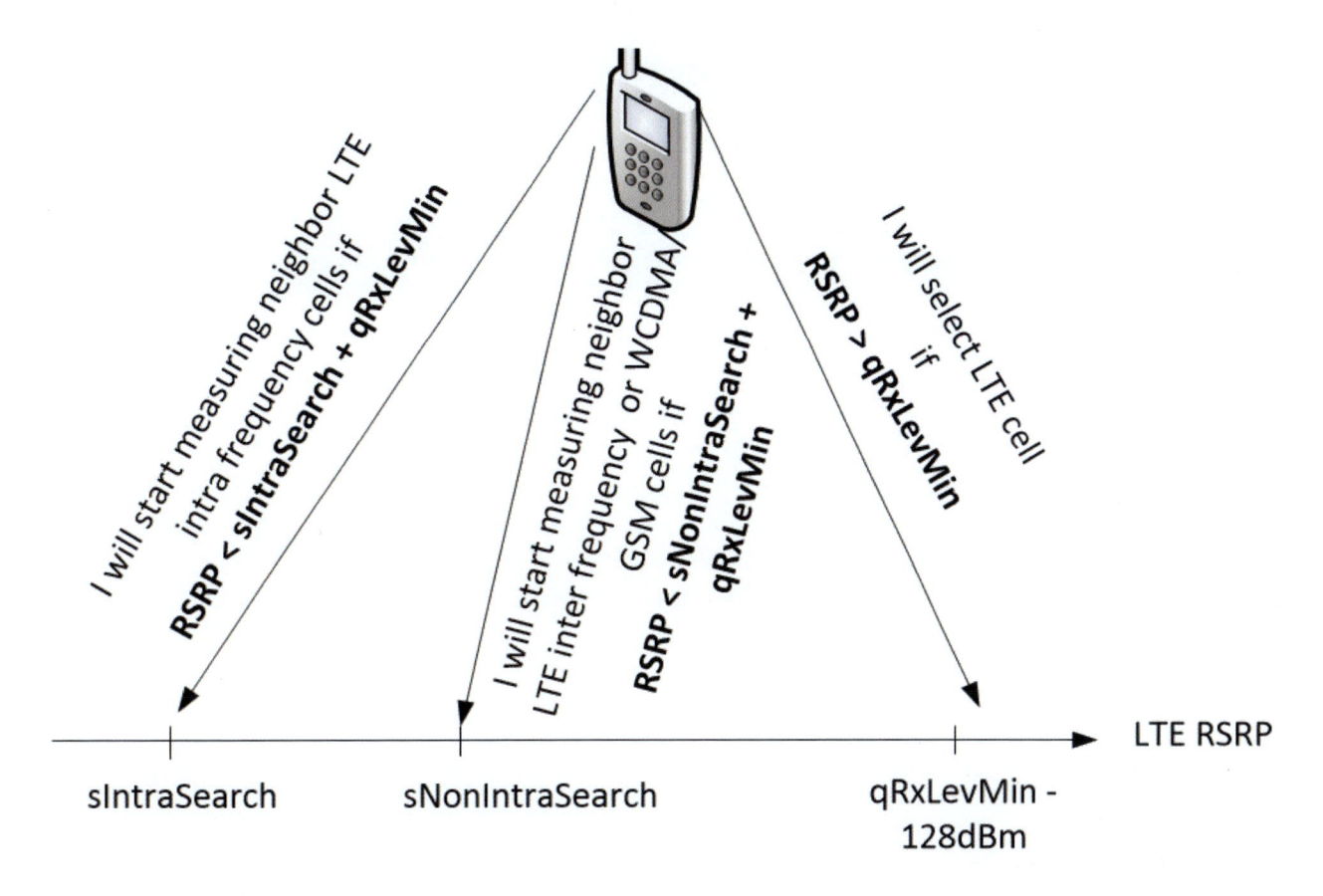

Figure 9.27 starting LTE cell reselection measurement to other RATs

For a phone to reselect to a suitable neighboring LTE cell, it must also satisfy cell selection criteria broadcasted in SIB1. Parameters configured in cell selection SIB1 are shown below

Table 9.17 LTE cell selection parameters

Parameter	Where will phone get this data?	Plan
QRxLevMin	LTE SIB1	-64 (actual is -128dBm)
QQualMin		-20

Cell reselection parameters to intra frequency and inter frequency LTE neighbor cells are broadcast in LTE SIB3. A phone is an LTE cell will reselect to another intra frequency same priority LTE cell if the following criteria is fulfilled for a specified timer called tReselection

$$target\ cell\ level = serving\ cell + qHyst$$

This means if target cell is better than the serving cell by some margin called hysteresis during a specified timer the phone will reselect the target intra frequency cell. Normally configured as 2 – 4dB

For cell reselection to inter frequency different priority LTE cell the story is a bit different. If for example a phone camped on an LTE cells wants to reselect low priority inter frequency LTE cell, the following criteria must be satisfied

serving cell RSRP $<$ ThresholdServingLow

And

target cell RSRP $>$ ThresholdXlow

When both of those condition are met for a specified timer parameter called EUTRA tReselection, the phone will reselect from the high priority serving LTE cell to the lower priority target LTE cell.

If now the phone wants to reselect back the high priority LTE inter frequency cell, the following threshold must be met for a specified tReselection timer

target cell RSRP $>$ ThresholdXhigh

A simple diagram to illustrate those thresholds is shown below

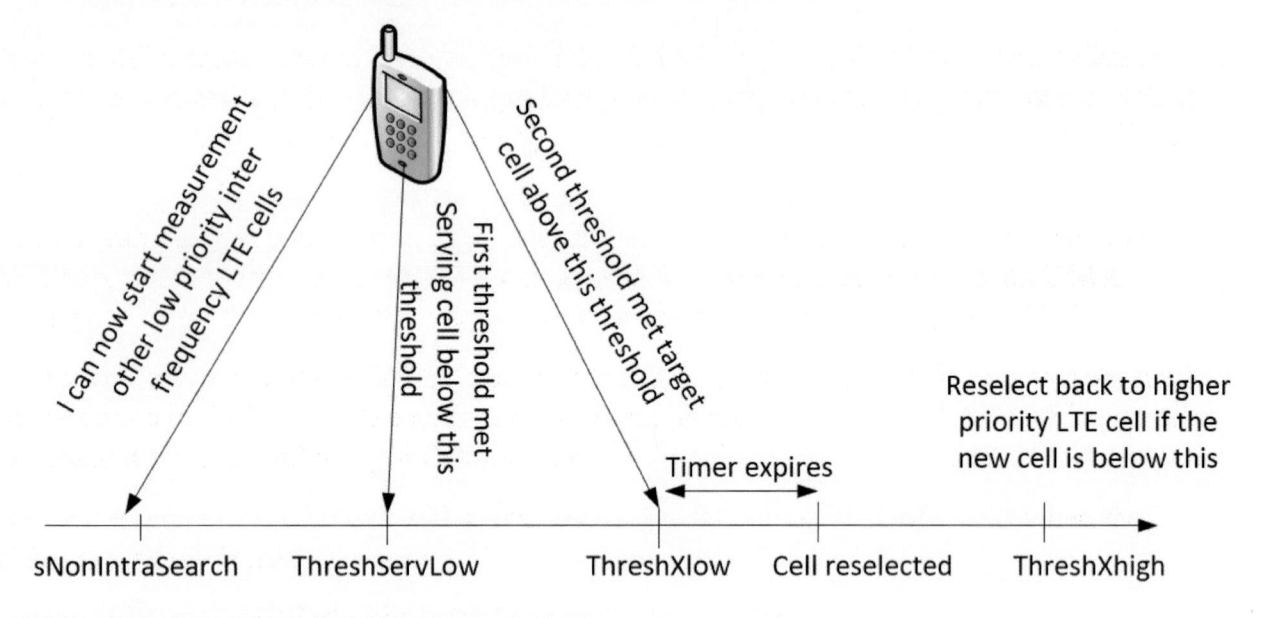

Figure 9.28 Priority based RAT reselection thresholds

As an example of LTE cell reselection planning, the following configuration is typical SIB3 content of high priority serving cell

Table 9.18 LTE intra frequency cell reselection parameters

Parameter	Plan	Comment
LocalCellId	0	
SNonIntraSearch config	CFG	
SNonIntraSearch	8 (-112)	Threshold for inter-frequency or Inter-RAT
SIntraSearch config	CFG	
SIntraSearch	20 (-88)	Threshold for intra-frequency measurement. Actual value is times two
QrxLevMin	-64	Minimum required RX level
ThrshServLow	6 (-116)	Must be less than sNonIntraSearch
CellReselPriority	7	
pMax	23	

The target low priority LTE cell configuration is also shown below

Table 9.19 LTE reselection to lower priority inter frequency parameters

Parameter	Inter frequency extra parameters	Plan
Local CI		0
CellReselPriority config	SIB5	CFG
CellReselPriority		6

ThreshXlow		10 (-108)
QrxLevMin		-64
EUTRAN tReselection		1s
DL EARFCN		255

Inter RAT cell reselection

Operators who own multi technology to serve their customers need to meet seamless interoperability of those technologies. In this section we will take a look at the following

- Reselection from high priority LTE to low priority WCDMA for coverage
- Reselection from low priority WCDMA to high priority LTE for better data service
- Reselection from high priority LTE to low priority GSM for coverage
- Reselection from low priority GSM to high priority LTE after for example circuit switched fallback (CSFB) call termination
- Reselection between WCDMA and GSM

These mobility must occur on time without delay or too early, and at the same time when cell reselection performance is not good, it results in bad user experience. Frequent cell reselection (ping pong) must be solved immediately by following troubleshooting techniques we will see later

To start with the first point above, users who are on LTE may need to reselect to a suitable WCDMA cell when for example CSFB call to WCDMA is ended or when the WCDMA coverage is better than the LTE coverage to take away the user from bad quality LTE.

To reselect to low priority WCDMA cell, the serving LTE cell broadcasts parameters required for the phone to do this job. The phone obtains these parameters form LTE SIB6 which contains information about neighboring WCDMA cells

The phone camped on LTE cell will start measuring neighboring WCDMA cells when the following criteria is met

$serving\ LTE\ cell\ RSRP < sNonIntraSearch + qRxLevMin$

Recall that we said the phone obtains sNonIntraSearch from SIB3 and qRxLevMin from SIB1

The phone will reselect to neighboring WCDMA cell when the following conditions are met

$serving\ cell\ RSRP < ThresholdServingLow$

And

target cell RSCP > ThresholdXlow

The typical configured LTE SIB6 by the RF planner is shown below

Table 9.20 LTE to UMTS reselection parameters

Parameter	Where will phone get this data?	Plan
UTRAN tReselection	SIB6	1s
LocalCellId		0
UARFCN		10855
CellReselPriority config		CFG
CellReselPriority		5
ThreshXLow		4 (-102) = –110 + (4)(2)
QRxLevMin (minimum RSCP of WCDMA)		-55 (actual is -110)

This means the phone will reselect to neighboring low priority WCDMA cell if the measured RSCP of the WCDMA cell is greater than ThreshXlow (RSCP), which is -102dB and the measured RSRP of the high priority serving cell is less than -116dBm (ThreshServLow)

Now the LTE call fallback to WCDM is over and the phone is now entering good LTE coverage, the phone must seamlessly reselect back to the LTE for better user experience

The phone will now using parameters broadcasted in WCDMA SIB19 reselect the LTE cell as follows. The SIB19 is sent by RNC to all WCDMA NodeBs. It contains LTE EARFCN and priority to be reselected.

Table 9.21 UMTS to LTE reselection SIB19 parameters

Parameter	Plan
Sib switch	SIB19
Serving cell Priority	5
EARFCN	1500
Neighbor cell Priority	7
ThreshXhigh	10 (must be greater than sNonItraSearch) = -108dBm (-128 + (10)(2)
QRxLevMin	-64 (actual value is -128)

NodeB version	R9
	Fast reselection switch enabled

This means whenever the LTE target cell threshold is greater than -108dBm reselect that LTE cell of the high priority. It does not matter if the serving WCDMA signal is better than the target LTE cell. All matters is LTE RSRP is greater than -108dBm (in SIB19)

A screenshot from WCDMA phone received SIB19 configuration is shown below. It was tested by running the SSID code ***#0011#** on Samsung Android device. The circled information "SIB 19 received" indicates that the operator in which this example phone is registered does allowed reselection from WCDMA to LTE by configuring SIB19 in the RNC and broadcasting it to WCDMA cells. That SIB19 messages contains LTE EARFCN.

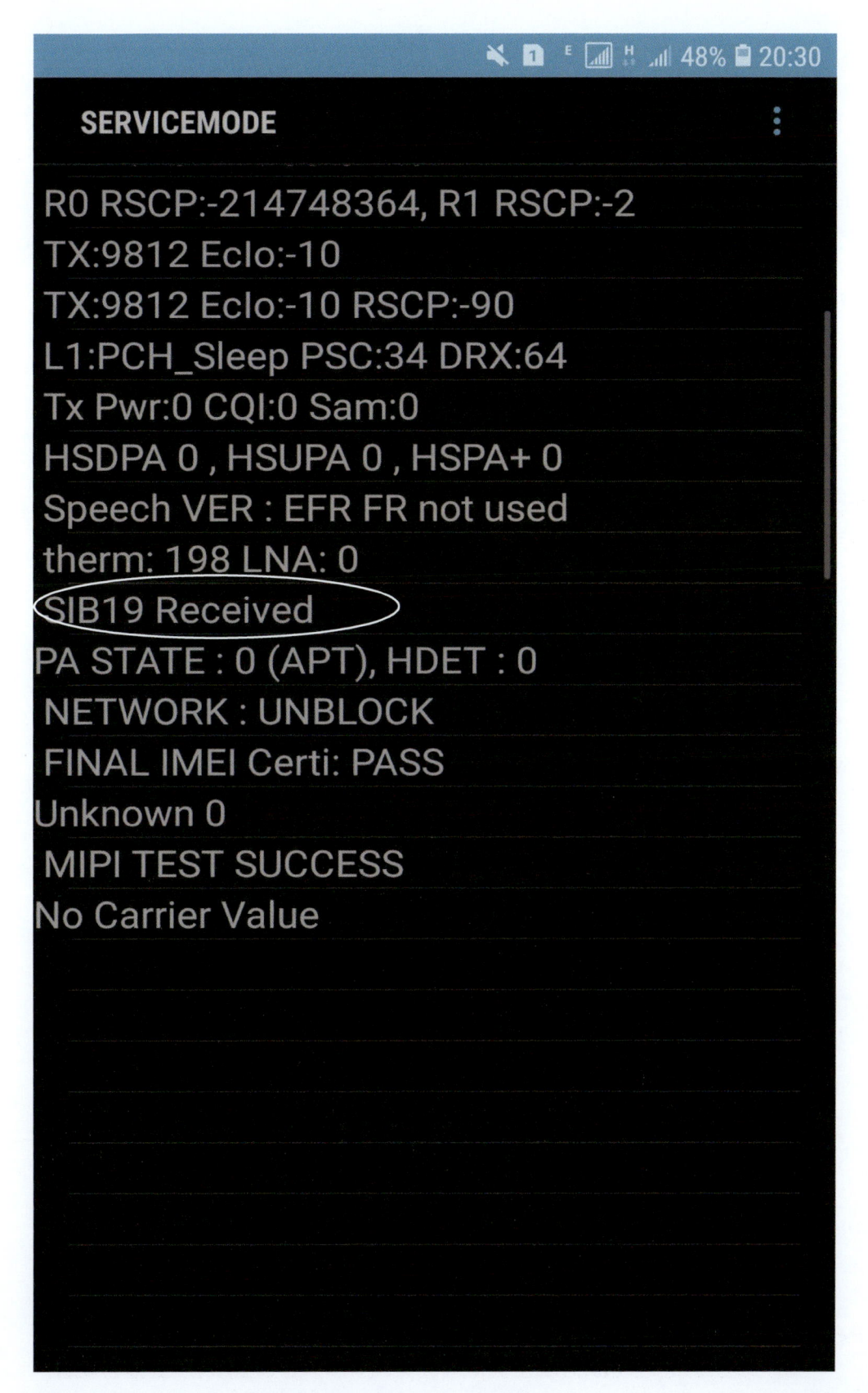

Figure 9.29 SIB 19 received on the phone from RNC

We end discussion on LTE and WCDMA interoperability in idle mode here. We now try to explain how reselection between LTE and GSM cells are planned and parameters involved along with typical parameters thresholds deployed in real networks

The idea is the same as that of LTE/WCDMA with the difference now being SIB6 is replaced by SIB7 broadcasted in LTE cell that contains GERAN or GSM reselection information.

A typical scenario of LTE to GSM for coverage based resection is summarized below

Table 9.22 LTE to GSM reselection parameters

Parameter	The LTE phone will read this SIB	Plan
ARFCN list		GSM-900 channels
LocalCellId		0
GERAN Treselection		1s
QrxLevMin		1 (-103) (2*1 − 105 = -103)
ThreshXLow	SIB7	4 (-95) (2*4 − 103 = -95)
CellReselPriority config		CFG
CellReselPriority		1
BcchGroupId		0

From this table, a phone on LTE serving cell will only reselect to low priority GSM cell when the target GSM cell RSSI is greater than -95dBm for specified timer configured in the BSC called GERAN Treselection. This condition is not only enough. The LTE RSRP must also be less than ThreshServLow (-116dBm) in SIB3 table discussed previously) for specified timer called EURAN tReselection

The corresponding parameters configured on the GSM cell is shown below. Remember you also need to create neighbor list configuration in conjunction with cell reselection parameters. Neighbor list configuration will need separate template file to be prepared.

Table 9.23 GSM to LTE reselection parameters

GSM Parameter	The phone will read in SI2quarter in GSM cell carried by BCCH	Plan
External LTE CELLID	These parameters configured on BSC and broadcasted on SI2quarter	To add an external LTE cell
External LTE cell name		
MCC,MCN,TAC,EARFCN,PCI		External LTE information
EUTRAN TYPE		FDD
LTECELLRESEL ENABLE		YES
GERANPRI		1
EUTRANPRI		7
ThreshXhigh		10 (2*10 – 128 = -108dBm)
EUTRANQRXLEVMIN		-64 (actual is -128)
Send2QuterFlag		Yes
Fast reselection switch		ON

This table indicates for a cell to reselect from low priority GSM cell to high priority LTE cell, the LTE measured RSRP by the phone must be greater than -108dBm

The final piece remaining before we end cell reselection interoperability between technologies is cell reselection between WCDMA and GSM

Not all operators have LTE technology. Due to low subscriber density and lack of enough capital investment, many operators provide only WCDMA and GSM to their customers. This was normal in the past. But recently internet service through WCDMA or HSPA is not sufficient for current users in urban centers. LTE is much need nowadays for video streaming and VOIP

The WCDMA cell is given high priority while the GSM cell low priority. The phone will always select the WCDMA unless certain thresholds are exceeded to reselect the GSM. The RF planner can control and tune the behavior of the cells.

A phone camped on WCDMA cell will start measuring neighboring suitable GSM cells (or even LTE if present) if the following condition is met

$EcNo \leq sSearchRAT + QqualMin$

Information about neighboring WCDMA cells is broadcasted to the phone via SIB11

For the serving WCDMA cell and target GSM cell the following parameters are configured on typical network deployment

Table 9.24 GSM WCDMA reselection parameters

Parameter	Plan	Node
Qqualmin	-16	RNC
QrxLevMin	-110	RNC
S searchRAT	2	RNC
FDDQmin	-14	BSC
QsearchI	7 (always) Always search WCDMA cells	BSC
FDD_RSCPmin	-105	BSC
FDD_Qoffset	0	BSC

This means using the above formula, for a phone camped on WCDMA to start measuring RSSI of neighboring GSM cells, the measured WCDMA quality level EcNo must be less than -14

$EcNo < 2 - 16 = -14 dB$

For a UE camped on low priority GSM to reselect high priority WCDMA cell, the following condition must be met. The other neighboring RATs are told via SI2quarter

- The measured CPICH *Ec/N*o value of the candidate WCDMA cell is equal to or greater than the parameter *FDDQmin*.

    Ec/No > FDDQmin

- The measured CPICH RSCP value of the candidate WCDMA cell is at least *FDD_Qoffset* [dB] better than the RLA (receive level average) of the serving cell and all suitable non-serving GSM cells.

    RSCP > RLA (serving, non-Serving GSM cells) + FDD_Qoffset

    • The measured CPICH RSCP value of the candidate WCDMA cell is equal to or greater than the optional parameter FDD_RSCPmin

    RSCP > FDD_RSCPmin

Intra RAT handover

We said in previous discussions, the phone will have signaling states. It can either be idle or dedicated. The exception is WCDMA where the dedicated state has four RRC states.

For phone in dedicated or connected mode and sending or receiving data, the network should provide seamless handover from one cell to next cell of the same or different RAT.

We address this section as follows

- LTE or EUTRAN intra RAT handover
- WCDMA or UTRAN intra RAT handover
- GSM or GERAN intra RAT handover
- LTE and WCDMA inter RAT handover
- LTE and GSM inter RAT handover
- WCDMA and GSM inter RAT handover

Intra RAT handover is handover between same technologies while inter RAT handover is handover between different technologies

To begin with LTE intra RAT handover, a phone in connected mode (call or data session) is moving from coverage of one eNodeB to another. The source cell and target cell must first have neighboring relationship planned and correctly configured. LTE and WCDM handover is based on event triggering while that of GSM is based on time measurement of the receive signal.

EUTRAN – EUTRAN handover can be further classified as

- Intra frequency handover
- Inter frequency handover

The handover thresholds are specified in events broadcasted to the phone. The phone then reports measurements after event thresholds are met by waiting a specified timer

Remember in connected mode, the phone had already established RRC connection with the EUTRAN. The EUTRAN sends RRC connection reconfiguration message in which the phone finds pre-defined event configuration by the RF planner

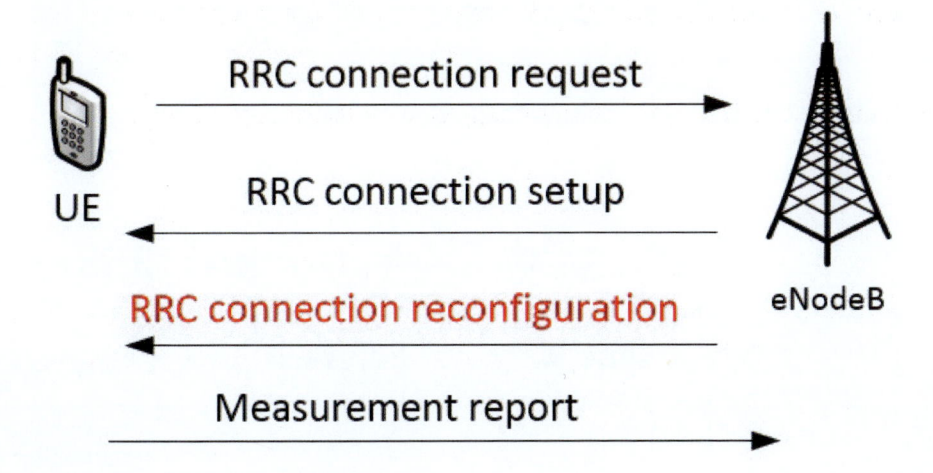

Figure 9.30 LTE events configuration and measurement reporting RRC messages

The events used to trigger LTE intra RAT (intra and inter frequency) handover are summarized in the table below along with an explanation

Table 9.25 LTE intra RAT handover events

Event name	Used for	Threshold diagram
A1	Serving cell is better than threshold 1. No need to trigger measurement report	RSRP vs Threshold 1
A2	Serving cell is worse than threshold 2. Now start measurement report after TTT expiry	RSRP vs Threshold 2
A3	when the neighbor LTE intra frequency cell becomes a certain offset better than the serving cell, the phone reports A3	Neighbor vs serving, Threshold2, Send A2, Send A3

182

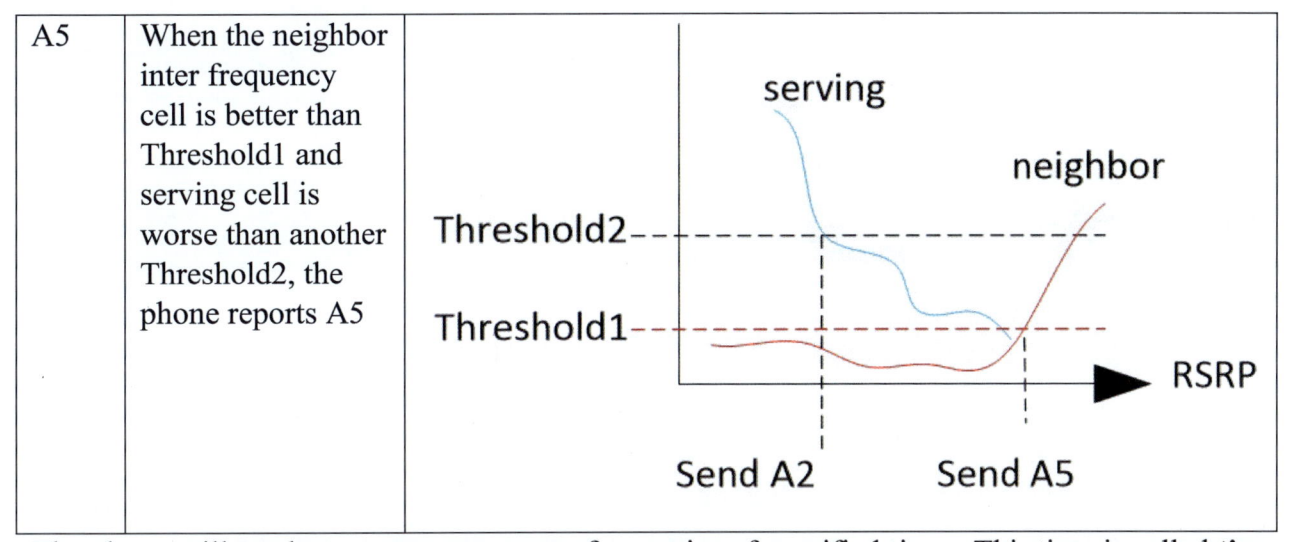

A5	When the neighbor inter frequency cell is better than Threshold1 and serving cell is worse than another Threshold2, the phone reports A5	*(diagram showing serving and neighbor RSRP curves with Threshold2 and Threshold1 lines, indicating Send A2 and Send A5 points)*

The phone will send measurement report after expiry of specified timer. This time is called **time to trigger (TTT)** and serves similar purpose as cell reselection timer. This means first the event is detected by the phone measurement. Then waits for TTT. When TTT expires, the phones will send the measurement report and the eNodeB will execute the handover, on X2 for example

A simple diagram to reinforce understanding A5 measurement report is shown below. In summary event A1 stops neighbor measurement, event A2 starts neighbor measurement, and events A3 and A5 measurement reports are sent to ask the eNodeB to initiate handover to intra frequency and inter frequency respectively.

Cellular network planning fundamentals

Serving — RSRP -112dBm
Neighbor — RSRP -105dBm

UE: *Wait a second! Neighbor is better than THD1 and serving is worse than THD2. I need to report A5 after TTT expiry*

RRC connection reconfiguration

A5 THD1 RSRP -106dBm
A5 THD2 RSRP -110dBm

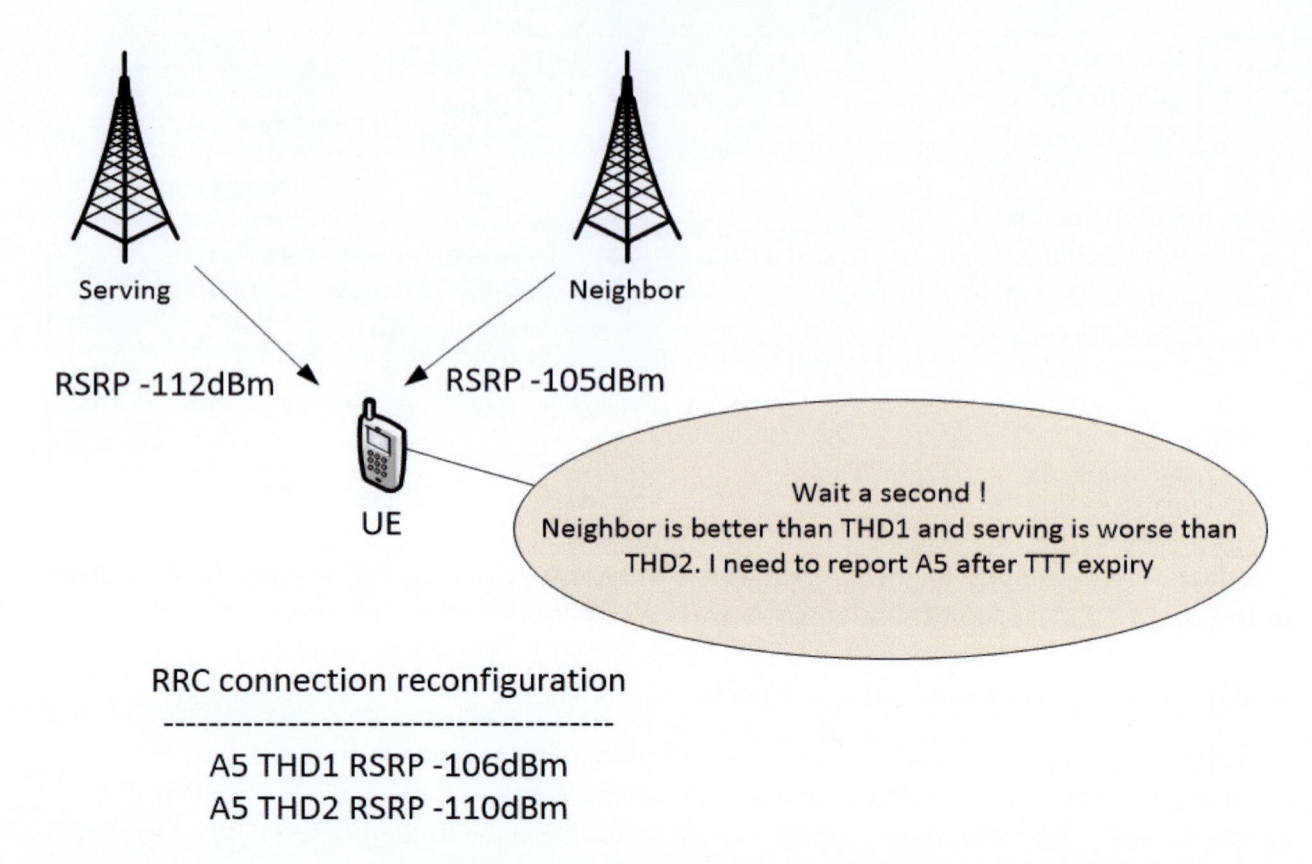

Figure 9.31 LTE inter frequency handover event A5 example

The events are delivered to the phone via **RRC connection reconfiguration** message. If you extract this message from the drive test tool, you should check that it matches with the RF plan. The parameters contained in the events configuration sent by the eNodeB is shown below

- Event ID
- Threshold
- Hysteresis (used to allow extra room for fast fluctuating signal)
- Time to trigger (time waited between event detection and reporting to eNodeB)

An example of a typical LTE intra handover parameters are shown in the table below

Table 9.26 typical LTE intra frequency handover parameter configuration

Parameter	Plan
LocalCellId	0
A3 Hysteresis	Default
A3 Offset	4
A3 TTT	240
Max Report Cell	Default

Report Amount	RSRP
A3 Trigger quantity	RSRP
Report Quantity	RSRP
Report Interval	120

An example of a typical inter frequency handover parameters are shown in the table below

Table 9.27 typical LTE inter frequency handover parameter configuration

Parameter	Plan
LocalCellId	0
A1 A2 Hysteresis	Default
A1 A2 TTT	Default
A1 Threshold RSRP	-102 (stop)
A2 Threshold RSRP	-108 (trigger)
A5 Hysteresis	Default
A5 Threshold1 RSRP	-106
A5 Threshold2 RSRP	-110
A5 TTT	480

LTE intra RAT handover can be S1-based or X2- based. In the latter, there is a link between source eNodeB and target eNodeB

Let us illustrate how intra RAT handover between UTRAN cells are planned. The events are quite different than those of EUTRAN but serve similar purpose. UTRAN intra frequency handover is based on soft handover in which new serving cell is established before removing the old cell. This is in contrast with LTE and GSM which use hard handover that require breaking the old cell before handing over to the new cell

A phone in UTRAN cell can simultaneously transmit to three cells called **active set.** Best cells are added to the **active set** and the next best cells stay in the **monitored set**. Cells that are strong but not included in the neighbor list will fill in the **detected set**

So when the phone is in UTRAN cell and in connected mode, it receives **measurement control message** from the RNC that contains events configuration. The phone will send back measurement report when the criteria in the events configuration are met

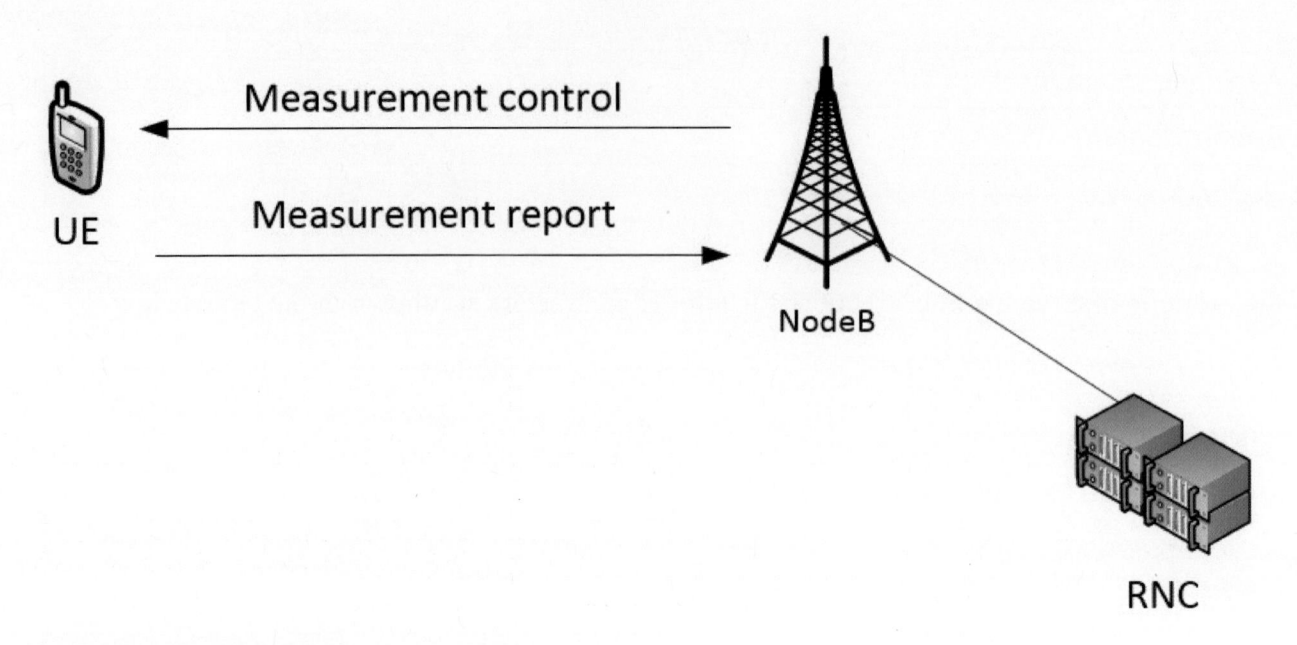

Figure 9.32 WCDMA handover commands

Events used in UTRAN intra frequency handover is summarized in the table 9.28 below along with diagram to explain it

Table 9.28 WCDMA intra RAT handover events

Event name	Used for	Diagram
1A	A cell in monitored set is strong enough and added to the active set	Monitor set cell crossing Threshold, Send 1A After TTT, axis CPICH
1B	A cell in active set is weak enough and should be removed from active set	Active set cell dropping below Threshold, Send 1B, axis CPICH
1D	Best cell added to the active set	Monitored set cell rising above Active set cell at Threshold, Send 1D, axis CPICH

187

An example of typical configuration of these events in real network is shown in the table 9.29 below

Table 9.29 typical WCDMA intra frequency handover parameter configuration

Parameter	Plan
Reporting range constant for event 1A	3
1A Hysteresis	0
1A TTT	60ms
1A reporting range	3.5dB
Reporting type	Event triggering
Measurement quantity	CPICH
1A amount of reporting	4
1A W	0
1A Threshold	-100dBm
1B reporting range	5dB
1B Hysteresis	0
1B TTT	640ms
1B W	0
1B Threshold	-105dBm
1D Hysteresis	4
1D TTT	320ms

Inter RAT handover

Now let us turn our attention to inter RAT handover planning between EUTRAN and UTRAN/GERAN. We start with EUTRAN and UTRAN

Most LTE network deployments today don't support Voice over LTE. Instead when a user initiates voice call inside the LTE cell, the call is handover over to the UTRAN or GERAN cells. This is process is called **circuit switched fallback (CSFB)**. Operators who want to carry voice traffic on the LTE network need to invest an IMS system which is expensive.

So on the LTE side handover between EUTRAN and UTRAN can be deployed as

- Circuit switched fallback using event B1
- Coverage based handover using event B2 (though B1 could be used)

- **Redirection** to UTRAN carrier in **RRC connection release** message

On the UMTS side

- Events 2D is used for starting compression mode and 2F for stopping it. Compression mode is activated during handover measurement. It is small time windows in which the carrier is not used for data transmission. It is used in cellular technologies as only transceiver is available to carry either measuring neighbors or cell data but not both.
- Event 3A is used for handing over to LTE

Cellular network planning fundamentals

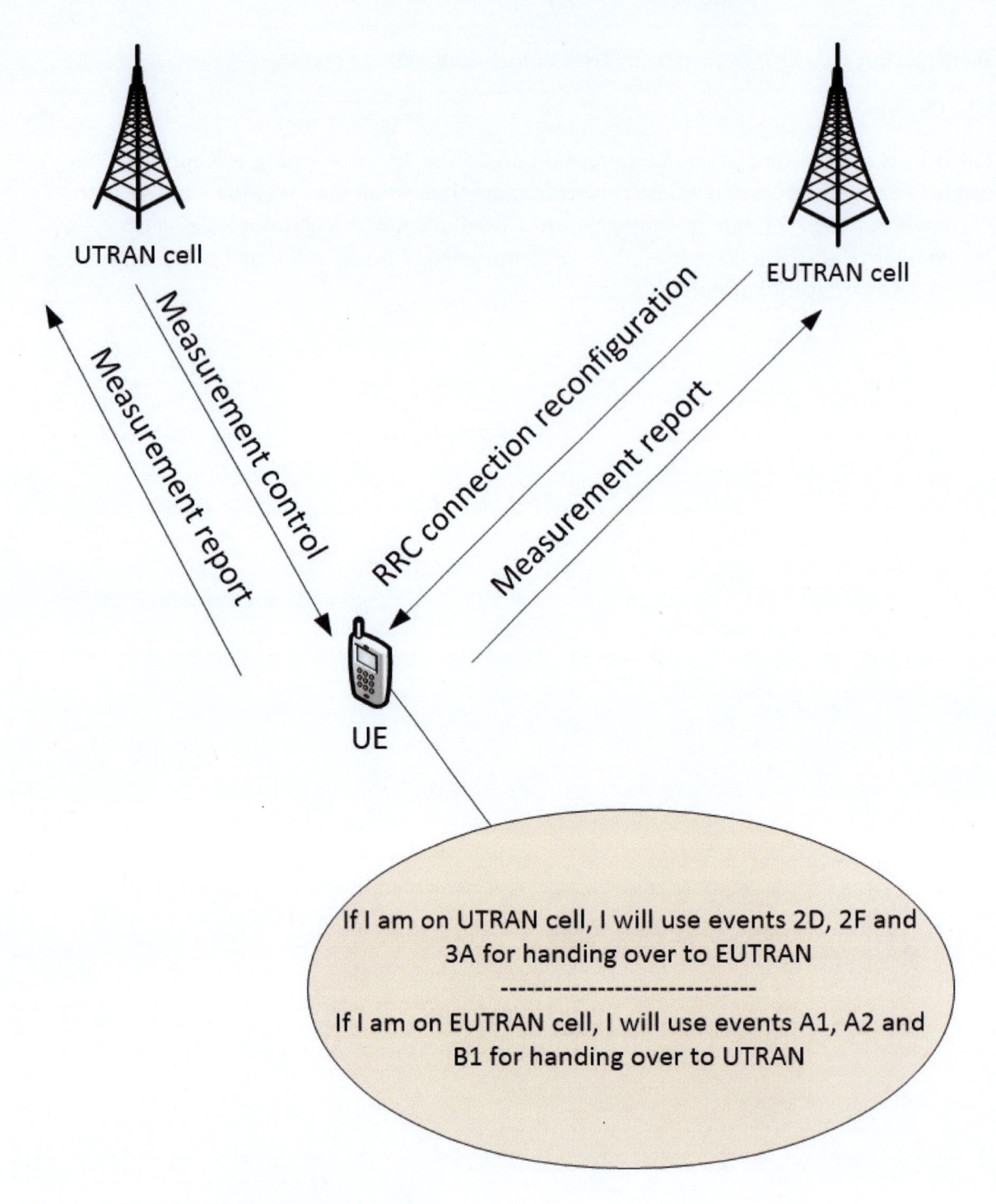

Figure 9.33 LTE WCDMA handover events

We can illustrate these events in graphs for easy understanding as shown below

Table 9.30 WCDMA inter RAT handover events

Even	Used for	Threshold diagram

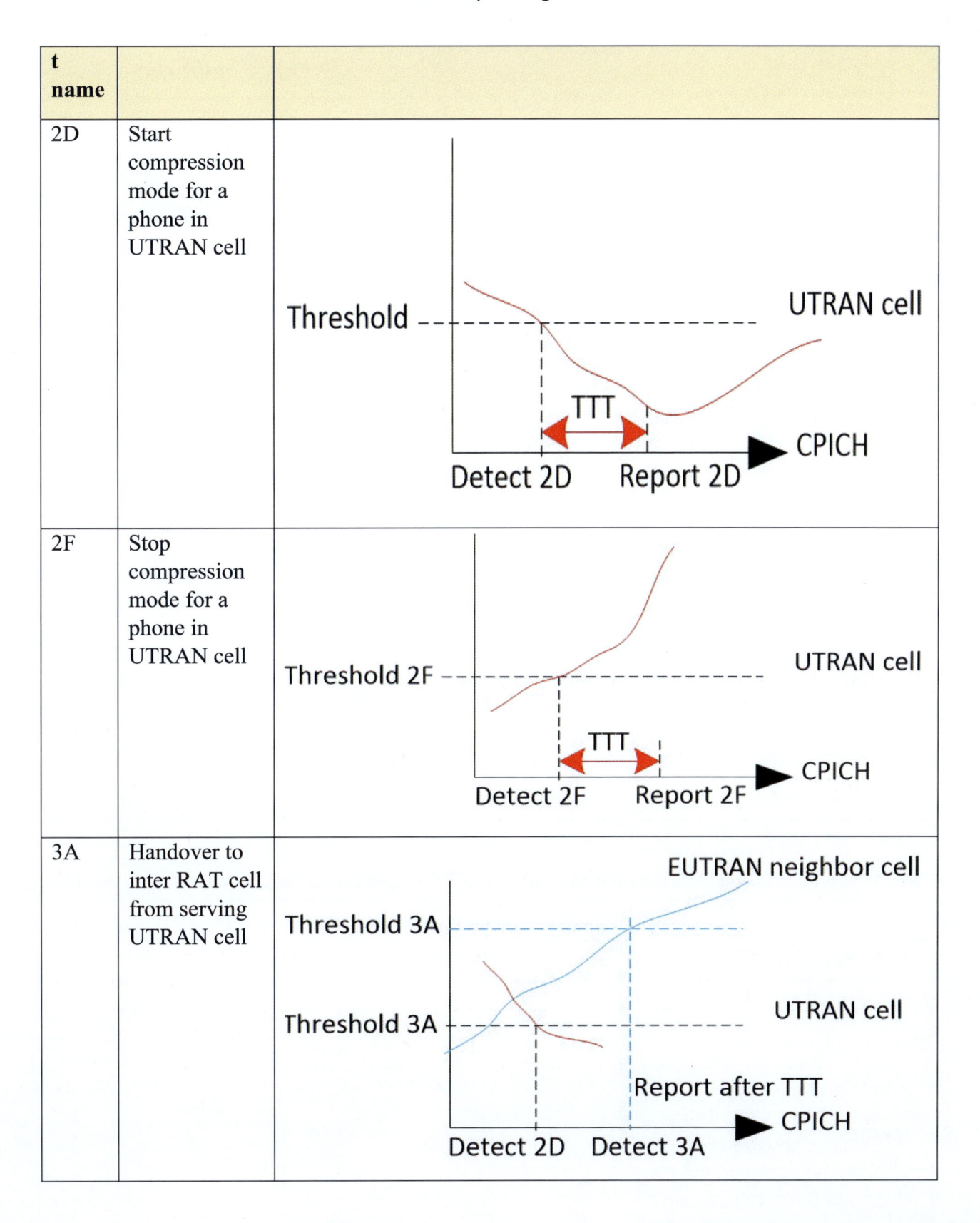

t name		
2D	Start compression mode for a phone in UTRAN cell	Threshold / TTT / Detect 2D / Report 2D / UTRAN cell / CPICH
2F	Stop compression mode for a phone in UTRAN cell	Threshold 2F / TTT / Detect 2F / Report 2F / UTRAN cell / CPICH
3A	Handover to inter RAT cell from serving UTRAN cell	Threshold 3A / Threshold 3A / Detect 2D / Detect 3A / Report after TTT / EUTRAN neighbor cell / UTRAN cell / CPICH

| B1 | Handover to inter RAT cell from EUTRAN cell | *(diagram: UTRAN neighbor cell curve rising above Threshold B1; EUTRAN cell curve falling through Threshold 2F; TTT interval between Detect A2 and Report A2; x-axis RSRP)* |

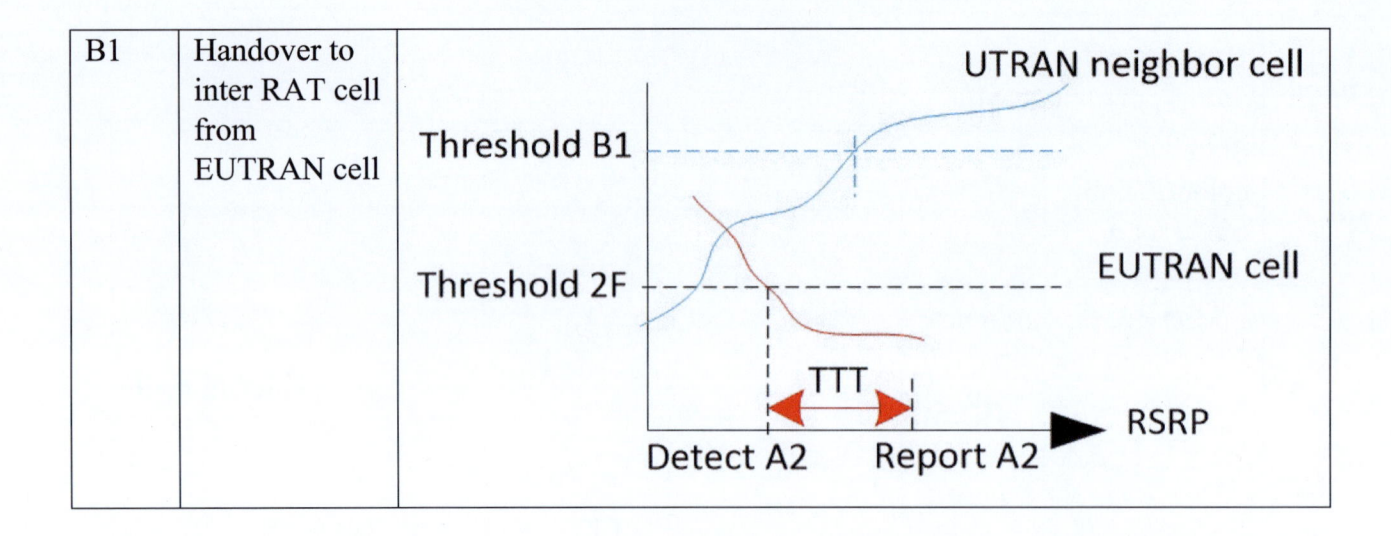

An example of typical handover configuration of EUTRAN and UTRAN cells is illustrated in the following table. Remember when using planning tool, different coverage and capacity need different handover thresholds

Table 9.31 typical LTE to WCDMA handover parameter configuration

Parameter (EUTRAN)	Plan
LocalCellId	0
A1 A2 Hysteresis	2
A1 A2 TTT	Default
A1 Threshold RSRP	-110
A2 Threshold RSRP	-116
B1 Threshold RSCP	-102
B1 Hysteresis	2
B1 TTT	Default
B1 Measured Quantity	RSCP
Blind handover priority	32

Table 9.32 typical WCDMA to LTE handover parameter configuration

Parameter	Plan	Note
Measurement quantity	CPICH-RSCP	

Threshold RSCP 3A	-95 dBm	Minimum strength of the serving UTRAN cell below which handover to LTE should be triggered. Too small ➔ handover to LTE delayed Too high ➔ UE in good UTRAN quality maybe unnecessarily handover over to LTE
W (Event 3A)	0	
Threshold RSRP 3A	-110	Minimum required LTE RSRP for handover to LTE
Hysteresis 3A	0	
TTT 3A	0	Period between detection of event 3A and sending of measurement report. If too large, event 3A triggering maybe delayed even when suitable GSM cell exists
2D Threshold RSCP	-103	Start WCDMA compression
2F Threshold RSCP	-99	Stop WCDMA compression

Similar parameter configuration on the UTRAN for WCDMA to GSM handover as shown below

Table 9.33 typical WCDMA to GSM handover parameter configuration

Parameter	Plan	Note
Measurement quantity	CPICH-RSCP	
Threshold RSCP 3A	-95 dBm	Minimum strength of the serving UTRAN cell below which handover to GSM should be triggered. Too small ➔ handover to GSM delayed

		Too high ➔ UE in good UTRAN quality maybe unnecessarily handover over to GSM
W (Event 3A)	0	
Threshold RSSI 3A	-95	Minimum required GSM RSSI for handover to GSM
Hysteresis 3A	0	
TTT 3A	0	Period between detection of event 3A and sending of measurement report. If too large, event 3A triggering maybe delayed even when suitable GSM cell exists

The corresponding parameters configured on the GSM cell is shown below

Table 9.34 typical GSM to WCDMA handover parameter configuration

Parameter	Plan	Note
3G_Search_Prio	1	Applies higher priority to UTRAN FDD measurements
FDD_REP_QUANTITY	RSCP	Coverage based handover based on signal strength
Qsearch_C	7	Always search 3G cells
FDD__Qoffset	0	Always select cell if applicable

Now let us give similar consideration to handover between EUTRAN and GERAN. Remember LTE without IMS infrastructure will not support circuit switched voice call. When a phone in LTE cell initiates normal voice call, the call will be taken care by the GSM circuit switched infrastructure. The mechanism that allows this process is called circuit switched call fallback (CSFB)

To make this process happen various interfaces between the LTE and GSM network must be physically connected and configured as shown below. One of them is SGi interface between the MSC and the MME that will carry combined attach to the MSC and MME by the phone as shown below

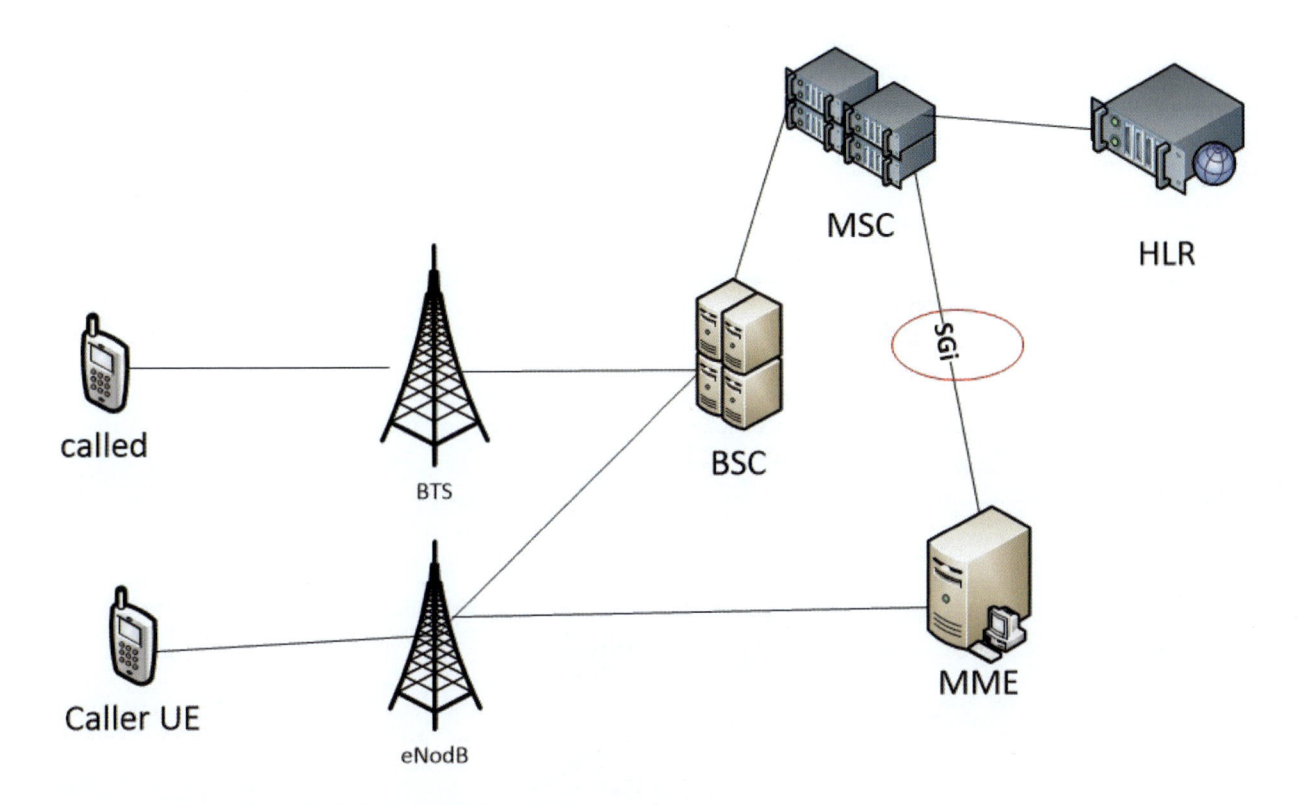

Figure 9.34 LTE to GSM CSFB interface

As we said earlier in this text, before configuring handover between inter RAT cells, first neighbor list should be prepared and configured. Handing over to a cell not in your neighbor list will fail and cause serious performance degradation on the quality of the service.

An example of GSM neighbor information to be configured on the LTE eNodeB is shown below

Table 9.35 example of LTE external GSM neighbor

Site name	Longitude	Latitude	MCC	MNC	CI	LAC	Band indicator	BCCH	NCC	BCC
A			637	81	10011	1	EGSM-900	1005	4	1
B			637	81	10012	1	EGSM-900	1007	4	3
C			637	81	10013	1	EGSM-900	1002	4	5

Once GSM external cell data is added to the eNodeB, the LTE external cell data is also added to the GSM BSC. Also LTE inter RAT handover uses event B1 as we seen in WCDMA as well in GSM

An example of typical event B1 configuration for LTE to GSM handover is given below

Table 9.36 typical LTE CSFB to LTE handover parameter configuration

Parameter	Plan
LocalCellId	0
B1 Hysteresis	1
B1 RSSI Threshold	-95
B1 TTT	640ms
A2 trigger quantity	RSRP

When the CSFB call completes, the phone goes into idle mode in GSM. Then idle mode cell reselection will bring the phone back to the LTE provided the LTE cell coverage is above certain threshold for specified timer. We covered this in the inter RAT cell reselection section.

Network optimization

After RF network plan is complete and sites launched, the next big phase that will continue forever and drain operator budget is RF network optimization.

The first step is defining what the operator management want? How do they want their network to be perceived? What performance target do they want to be achieved?

For example a certain operator may want the following performance targets for the RATs

Table 9.37 typical cellular operator performance objectives

Performance objective	Target
Coverage	> 95%
Quality	> 95%
Call drop rate	< 2%
Handover success rate	> 98%
Accessibility success rate	> 98%
Data session drop rate	< 2%
Voice call speech quality index (SQI)	> 20

Cellular network planning fundamentals

3G user throughput	As per acceptance test
4G user throughput	As per acceptance test
3G/4G latency	< 50ms

The RF drive test engineer is required to have in depth understanding of various network issues and KPIs often seen in new and operational network and how to troubleshooting using the following tools

- Drive test software and post-analyzer installed on dedicated laptop
- RF planning template sheets
- Cell file preparation and import to the tool as well as maps
- GSP mounted on the drive test vehicle
- Route planning with perfect timing and car speed is reasonably slow
- Test phones
- FTP server for 3G/4G internet test
- Compass and channel scanner

At typical setup of the drive test tool is illustrated below

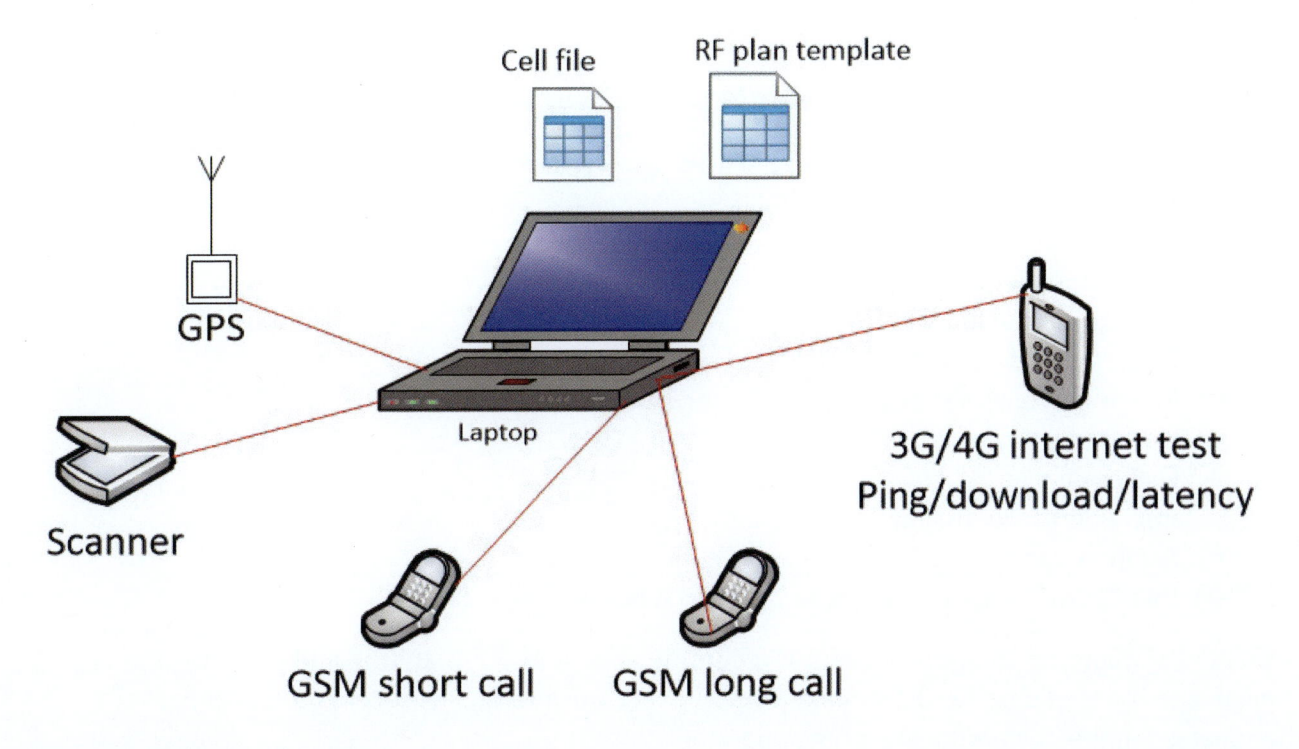

Figure 9.35 RF drive test tools

The following figure gives an overview of different types of KPI network that need to be understood in greater detail once network enters the operational phase

197

Cellular network planning fundamentals

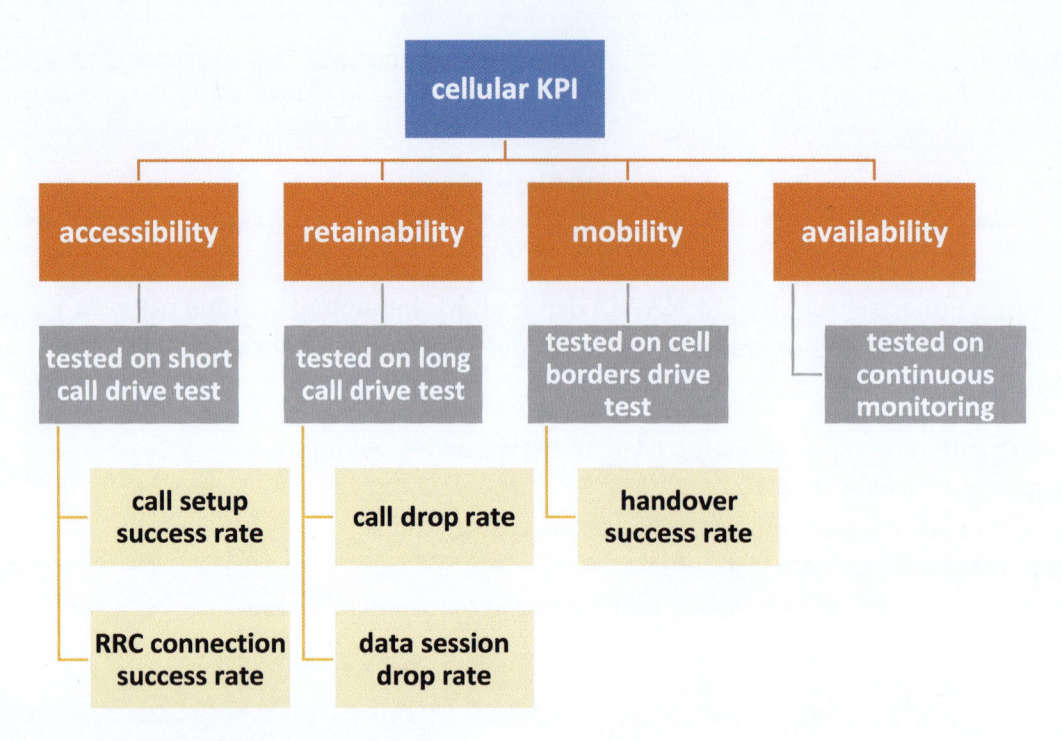

Figure 9.36 RF drive test basic KPI examples

The underlying causes of observed poor network KPIs are mainly RF with some others related to networking and core network

Some of the RF issues that result in poor network KPI include

- Interference
- Weak coverage and coverage holes
- Unbalanced uplink and downlink
- Weak transmit power
- Excessive antenna tilt
- Ping pong cell reselection and handover
- Missing neighbors
- Bad frequency plan
- Bad code plan
- PSC and PCI collision
- Sector swap
- Installation malpractice due to high VSWR

So the RF planning and optimization is a continuous cycle. RF planning and optimization is headache if not planned and executed properly. The air interface is where the customer is connected and is the reputation of the operator

Let us take few examples and cases. Consider while doing drive test, it was observed high percentage of session drop. Then after drive test analysis, it was observed ping pong (frequent) handover from one cell to another was evident in the session drop area as shown below

Cellular network planning fundamentals

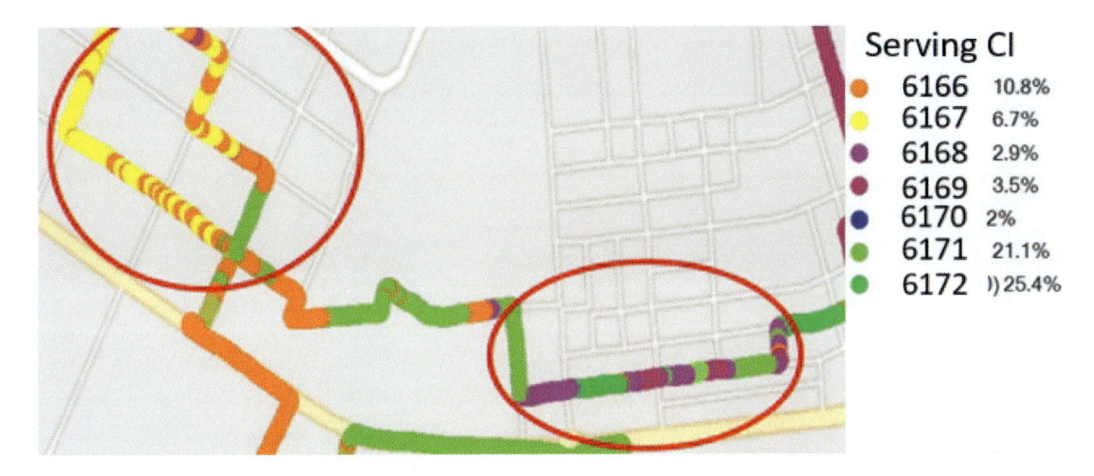

Figure 9.37 GSM drive test ping pong handover example

This ping pong handover could be cleared by tuning handover parameters (thresholds, TTT, hysteresis) and physical adjustment of antenna azimuth and tilt. The RF planning tool does have cell optimization module, that can give best azimuth, tilt and antenna height if done in the right way.

As a second example, suppose new LTE cell was deployed in area where there is high HSPA load as reported by drive test engineer. Now the RF planner develops and configures RF parameters for the cell. Everything is setup and cell connected to the core network by microwave

Now when the cell was turned on, it was observed all phones could not latch on the new LTE cell. Only drive test pocket phone can latch when locked with the LTE carrier. What is issue was asked by the management?

Figure 9.38 phone cannot camp on newly installed LTE cell

After careful investigation by the RF planner, he found out that the cell plan template and what was configured on the cell are totally different. He read SIB and MIB messages to understand cell configuration. For example he found out that RS power was too small and the carrier was incorrect. Several other parameters were also set incorrectly such as PCI, TAC, and the cell selection parameters. There was also high interference in the cell.

In summary the RF planner took the following steps to solve "phone cannot camp on new LTE cell" issue in the figure below

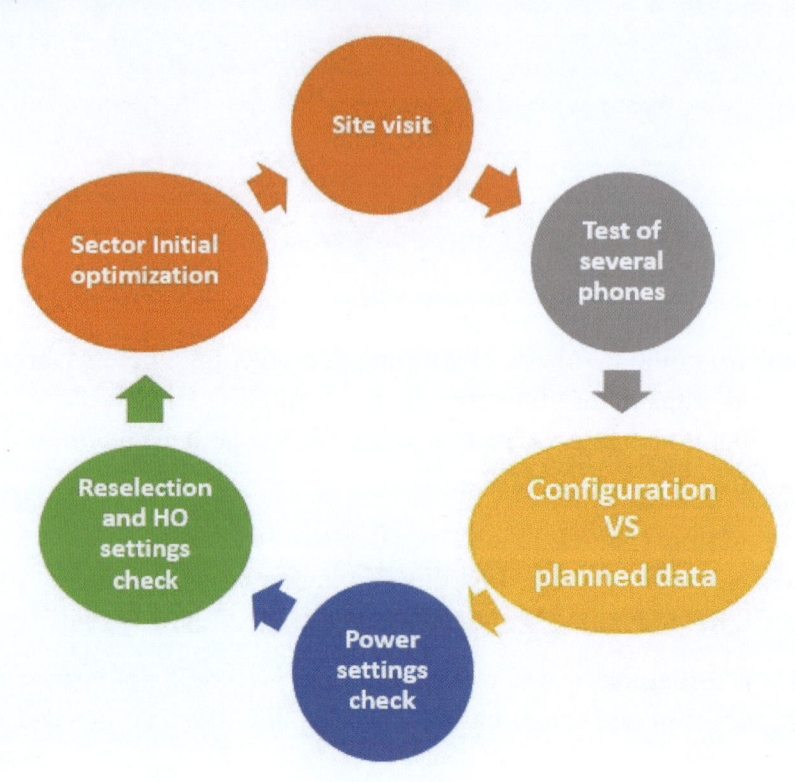

Figure 9.39 Troubleshooting RF plan and cell configuration

Let us discuss several other optimization cases as follows

Case I: weak coverage

Weak coverage can come into different forms and are explained in the following table

Table 9.38 weak coverage classifications

Coverage issue	Cause	Solution
DL coverage hole	Distance between two cells is large enough, hole emerges between them	Add new cell
UL high RSSI	Excessive down tilt Phone transmit power saturates to maximum	Tune UL power control parameters Adjust antenna tilt
UL/DL imbalance	UL coverage less than DL	Adjust antenna tilt

	coverage Other system alarm	Optimize UL and DL coverage in the planning tool
Overshooting	A cell of excessive height pollutes territory of other cells. It results weak coverage of the over shooter cell	Reduce antenna height Apply antenna tilt Reduce power
Lack of dominant best server	Several equally strong cells present at one location. The phone then will go ping pong cell reselection and handover	Reduce power Introduce new cell that over powers the rest

As a case example weak coverage of WCDMA cell is given below

Table 9.39 example of weak coverage WCDMA case

Problem	Definition	solution
Weak coverage	CPICH RSCP < -95dbm	- Increase TX power or antenna gain - Adjust antenna tilt and azimuth - Increase antenna height - Add new NodeB
Overshoot coverage	Coverage spots of NodeB 1 forms isolated islands in coverage area of NodeB 2	- Adjust antenna tilt and azimuth - Add the overshooting cell as neighbor list to victim cell to avoid missing neighbor - Change site location - Tune cell reselection parameters to discourage UEs from selecting that cell - Lower antenna height
Unbalanced coverage between uplink and downlink	Uplink coverage peaks maximum from UE transmit power	- Check uplink interference from RTWP in the RNC - Check NodeB alarms such as power failure
Lack of dominant best server	No best server in area. Ping pong handover results	Increase best signal coverage and reduce other weak signals in the

	which increases call drop probability	affected area

Some of the cause of WCDMA weak coverage include improper parameter settings, improper site location, interference, and bad installation practices

A typical optimization process for WCDMA networks is summarized in the following diagram below

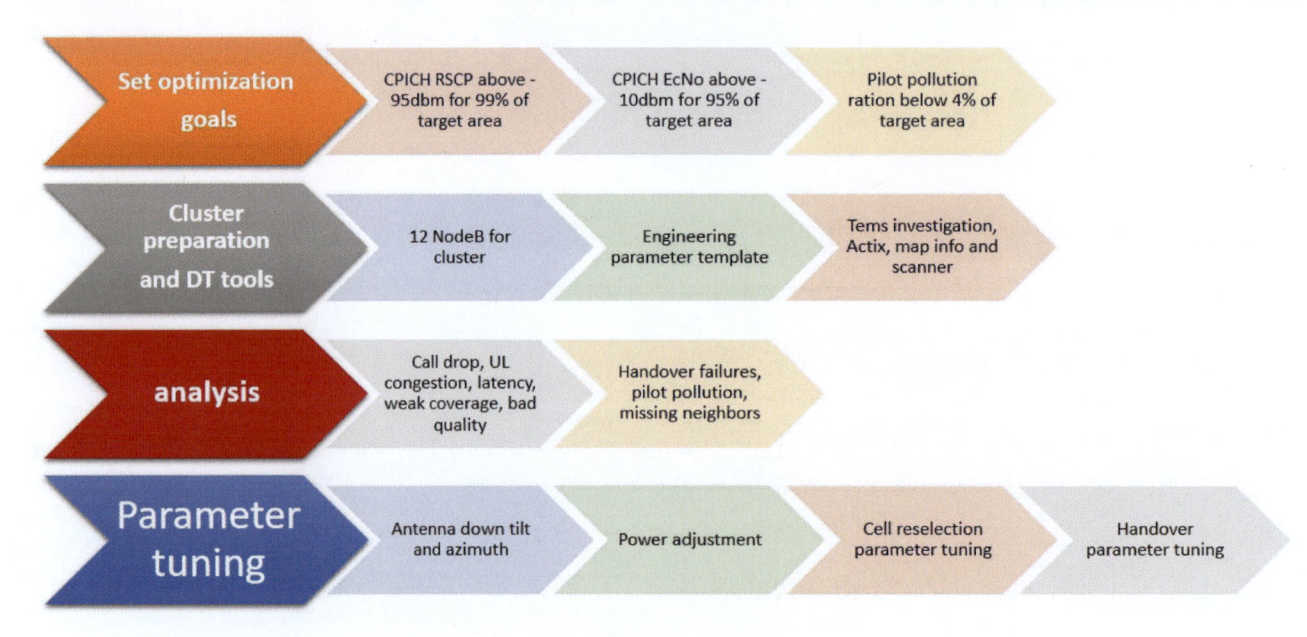

Set optimization goals	CPICH RSCP above -95dbm for 99% of target area	CPICH EcNo above -10dbm for 95% of target area	Pilot pollution ration below 4% of target area
Cluster preparation and DT tools	12 NodeB for cluster	Engineering parameter template	Tems investigation, Actix, map info and scanner
analysis	Call drop, UL congestion, latency, weak coverage, bad quality	Handover failures, pilot pollution, missing neighbors	
Parameter tuning	Antenna down tilt and azimuth	Power adjustment	Cell reselection parameter tuning	Handover parameter tuning

Case 2: Cause of WCDMA high CS call drop rate (CDR)

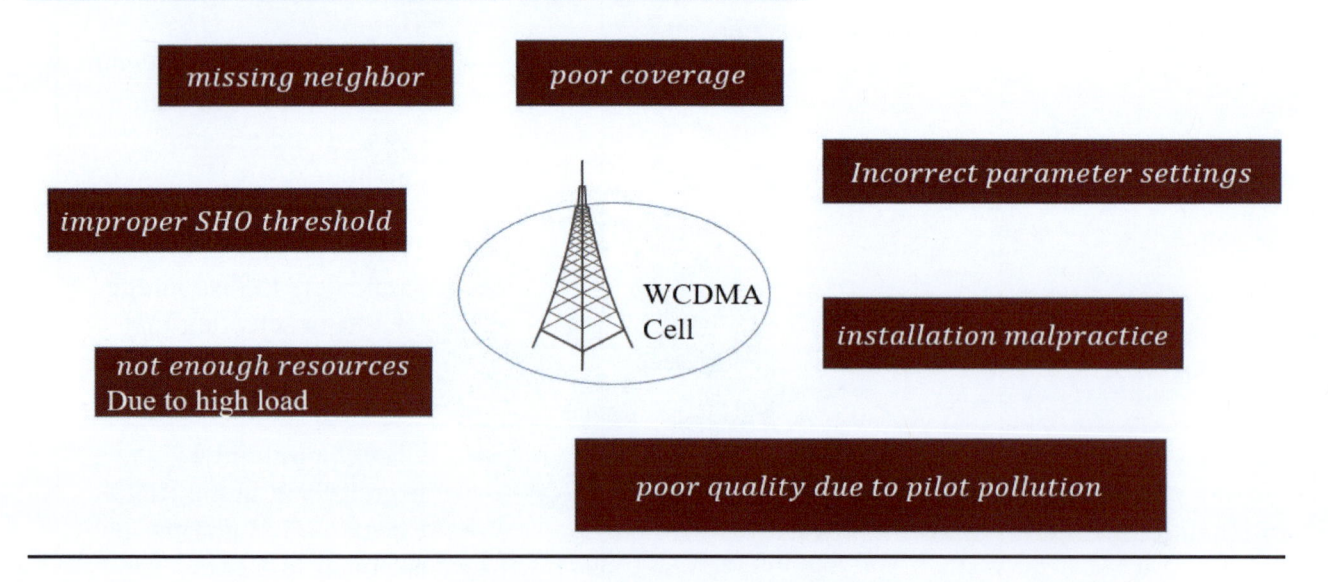

Causes around a WCDMA Cell:
- *missing neighbor*
- *poor coverage*
- *improper SHO threshold*
- *Incorrect parameter settings*
- *not enough resources Due to high load*
- *installation malpractice*
- *poor quality due to pilot pollution*

Case 3: Improving WCDMA call setup and AMR CS voice quality

Access takes place on RACH in which UE sets preamble initial transmit power

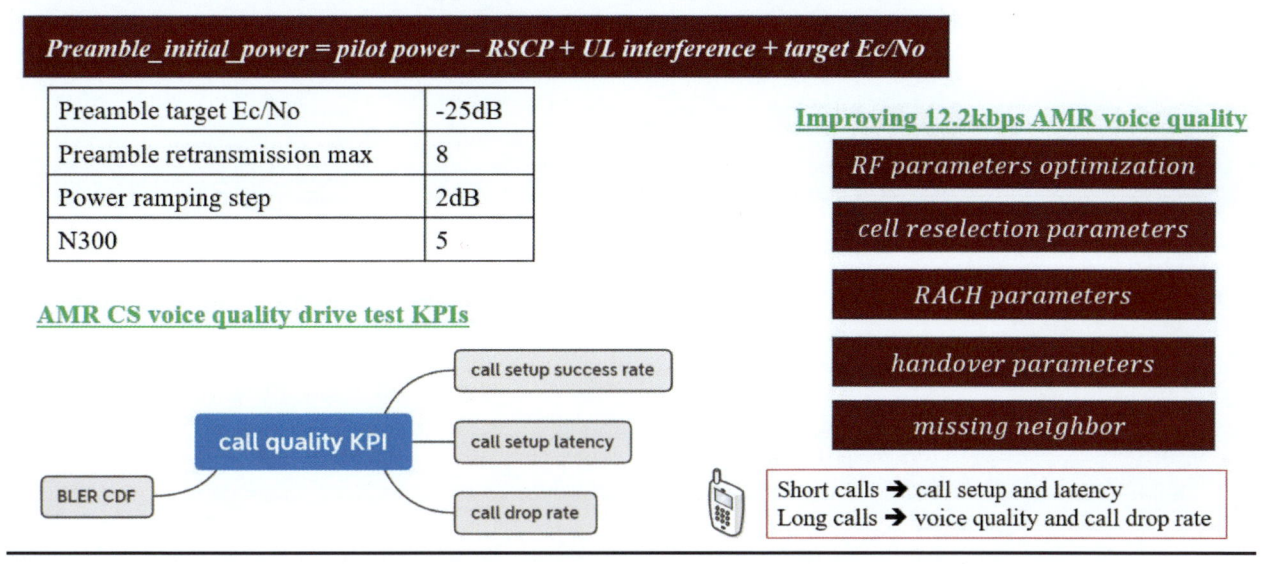

Preamble_initial_power = pilot power − RSCP + UL interference + target Ec/No

Preamble target Ec/No	-25dB
Preamble retransmission max	8
Power ramping step	2dB
N300	5

AMR CS voice quality drive test KPIs

call quality KPI:
- BLER CDF
- call setup success rate
- call setup latency
- call drop rate

Improving 12.2kbps AMR voice quality
- *RF parameters optimization*
- *cell reselection parameters*
- *RACH parameters*
- *handover parameters*
- *missing neighbor*

Short calls → call setup and latency
Long calls → voice quality and call drop rate

Case 4: GSM call quality degradation and voice crack

In weak coverage issues discussed above, the user on GSM call may arrive at the coverage hole or where there is overshooting signal coming from far cell not in good UL synchronization. The user may then hear cracked voice of the other party. If TC frequency plan is done incorrectly, interference may also arise. Normally hopping reduces interference in GSM networks. Voice cracks could also be caused by bad voice encoder in the BTS system. It could also be cause by UL and DL coverage imbalance due to high antenna height and high tilt or bad location of the site on the roof. Some ways to improve voice quality is to use enhanced full rate that encodes the voice at 12.2kbps, reducing cell power of overshooting cells, good frequency plan, good neighbor plan, checking installation VSWR, checking installation sector swap and also microwave backhaul transmission. Issues on core network A-interface (such as switching loop incase the connection is Ethernet) may also cause bad voice quality and call drops. Mismatching core network licenses may also cause problems

Poor voice call quality could be checked on drive test by

- Checking RxQual
- Checking SQI
- Check BER / FER
- Checking coverage

Case 5: GSM access failure can cause blocked calls

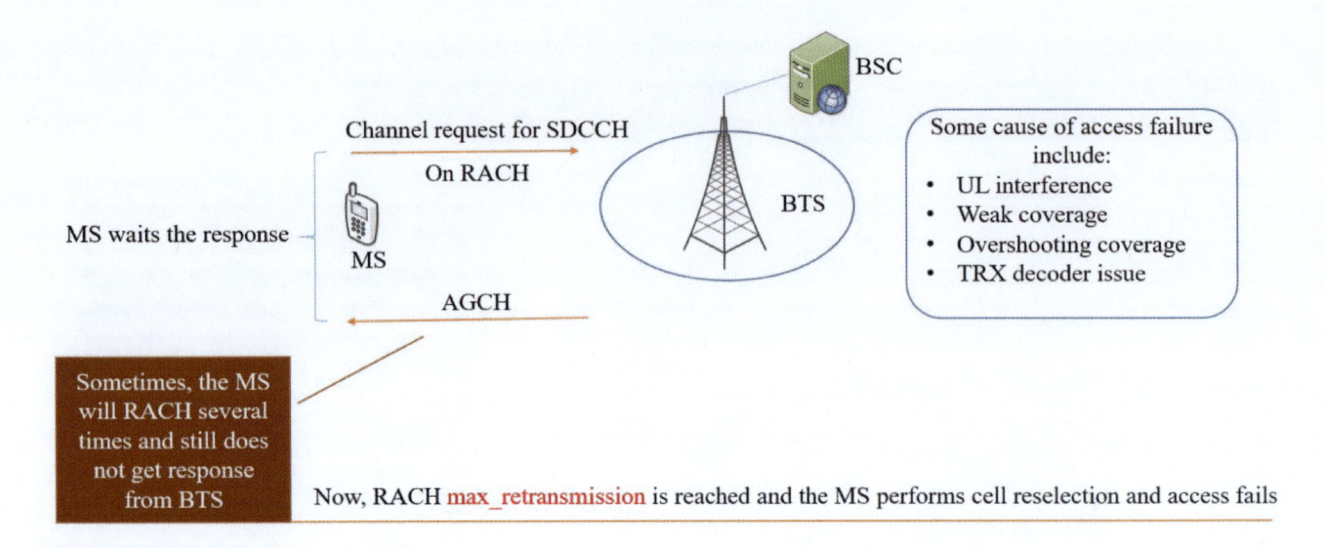

Figure 9.40 GSM access failure

Case 6: **HSPA signal is very weak for users located close to the cell and internet is very slow**

Internet speed on HSPA cell can be caused by various issues including networking, RF, installation or even power problem. We will take an RF example observed on the sector serving these users when drive test was conducted and analyzed

It was observed the sector serving these users located at cell center was very high in height and even excessive antenna down tilt didn't helped focus the signal. The drive test snapshots are shown below and could solved by lowering antenna height and adjust its tilt

The RSCP was too weak for this area (< -115dBm) as shown below

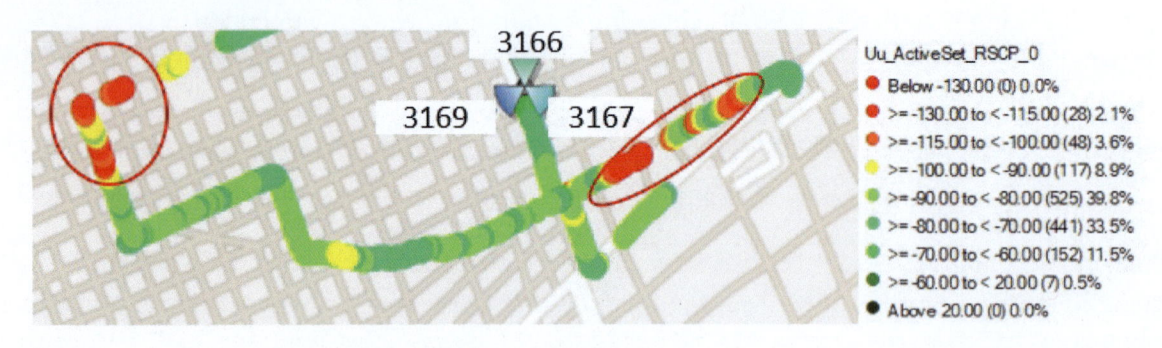

Figure 9.41 Drive test case of weak HSPA coverage

The EcNo was also very poor (< -15dBm)

Figure 9.42 Drive test case of poor quality HSPA cell

Case 7: high percentage of call drop

Call drop is one thing customers get annoyed. It is very bad user experience and if not solved urgently by the operator it may force even loyalty customers to discard that operator simcard.

Call drop can be caused by any or combination of the following RF issues

- Weak coverage
- Interference
- Missing neighbor which is strong
- System RF timers not tuned
- Two adjacent cells using carrier or code plan
- Sector swap
- VSWR alarm

Case 8: pilot pollution has hindered HSPA speed, operators are running at high operational cost

Pilot pollution occurs as a result of many different servers at one place without anyone being dominant to others. Typical areas where pilot pollution is prevalent is illustrated in the following

Tall buildings   Wide roads   Road junctions

Figure 9.43 common areas of HSPA pilot pollution

Pilot pollution can also be caused by overshooting cells which are strong enough. If the active set size increase beyond 3, it can create interference and missing neighbor issues.

Pilot pollution can be characterized by the high RSCP, quality level less than -15dBm, high downlink BLER, bad user experience when browsing the net, and high CDR due to ping-pong handover. In fact if you see on drive test any of the signs in the below table, then expect HSPA pilot pollution

Table 9.40 signs of WCDMA pilot pollution

EcNo	< -15dBm
RSCP	> -90dBm
BLER	> 10%
Capacity	Low due to bad quality
Call drop	Prevalent due to frequent handover

It is therefore recommended for operators to plan their networks using sophisticated planning tool coupled with human experience and judgment. Unfortunately, most operators opt for increasing system capacity at high cost, installing more sites without looking at below the bar and solve on what is really biting on their network. They could still increase speed and capacity by focusing on their existing infrastructure and fighting against pilot pollution and other noise effect and hence save large amount of money.

The drive test engineer may try the following to at least combat pilot pollution

- Antenna azimuth and tilt proper adjustment
- Pilot power adjustment of the interfering signals
- Use micro cells in places with multiple dominant signals

Case 9: too many HSPA session drops at cell border

The drive test engineer carries out an investigation. He uses one test phone in dedicated mode HSPA voice call. As soon as he enters cell border session drops. In fact there is high percentage of session drop he records and checks on the post-drive test analyzer

After analysis he concludes that the cause of the session drop was missing neighbor. A cell in the monitored set was tried to be handed over to, but because that cells was missing neighbor, the handover fails and the session drops. He could see the signal strength of the missing cell in the drive test tool neighbor measurement table.

A diagram illustrating missing neighbor and its effect is shown below

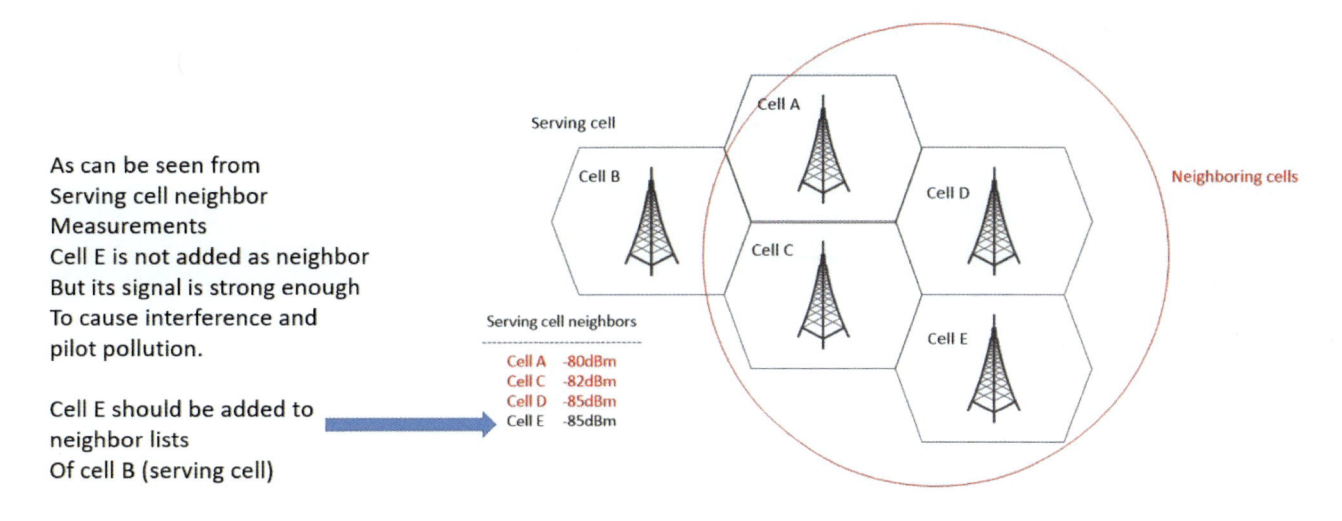

As can be seen from
Serving cell neighbor
Measurements
Cell E is not added as neighbor
But its signal is strong enough
To cause interference and
pilot pollution.

Cell E should be added to
neighbor lists
Of cell B (serving cell)

Serving cell neighbors
Cell A   -80dBm
Cell C   -82dBm
Cell D   -85dBm
Cell E   -85dBm

Figure 9.44 missing neighbor example

Case 10: the phone always latches on GSM instead of LTE even close to cell, user are tired of using flight mode to bring back the LTE. This is not seamless user experience

The author has seen this in many times and helped some operators tackle it. This happens in various ways. One way is as the user is holding his phone in idle mode, he always sees E instead of 4G and while in dedicated mode and watching YouTube, the speed suddenly collapses and all he sees now is E on top of his phone bar. The other way is as the user makes voice call on LTE, the call is redirected to GSM. When the call terminates, the phone stays in GSM forever unless manual flight mode is used to bring LTE back

The solution to all these problems converges into cell reselection and handover parameter settings

- If reselection threshold to LTE is set to a high value (> -90dBm), then the phone will stay with GSM, as at LTE cell border and indoor signal level is always less than -90dBm. Hence ***thresholdXhigh*** configured in GSM cell should be reasonable to allow the phone to always go to LTE. The threshold to leave LTE (***thresholdServLow***) should be set low value while keeping in mind quality
- If cell reselection parameter ***tReselection*** in LTE is set to very small, then the phone will quickly go to GSM. Large hysteresis could be applied to compensate for fast fading of signal fluctuation
- In connected mode phone on LTE, event B1 **TTT** should be increased to delay the handover to LTE in data transfer session. In CSFB call termination, again cell reselection parameters should be adjusted as indicated in the first bullet point

Case 11: LTE UL throughput is very small (difficult to upload small file or watch live video streaming)

Whenever users complain of low upload or live streaming on LTE connection, the drive test engineer will understand they mean UL throughput issue. The drive test tool is checked especially UL RSSI which indicates interference. A high UL RSSI means high UL interference and hence bad UL throughput. UL PUSCH BLER percentage another good indicator of UL quality status. Sometimes there could also be insufficient UL power, and thus UL power control parameters should be adjusted. An example of LTE power control parameters of typical network is given in table 8.41 below

Table 9.41 typical LTE UL power control parameters

Parameter	Description	Plan
PoNominalPUCCH	Nominal power for UE PUCCH TX power calculation	-97
PoNominalPUSCH	Nominal power for UE PUSCH TX power calculation	-104
A	Indicates the compensation factor for path loss.	7

Please notice that in the above case, DL issues can also introduce UL quality issues. The drive test engineer is required to also test DL throughput in conjunction with the UL throughput. If there is bad coverage in LTE DL, the PDSCH BLER will increase more than 10% and reduce the DL throughput significantly. If there is interference in LTE DL, the UE will report bad CQI report (<15) and in turn get lower throughput QPSK form the eNodeB packet scheduler.

An example of LTE DL bad coverage is shown in figure 9.45 and 9.46. Notice that the RSRQ is poor outdoor (< -100dBm) while RSRQ is very good (> 5_dB). The worst part is that this network was unloaded when this drive test was taken. Once users are placed on this cell, then RSRQ will significantly reduce as interference increases

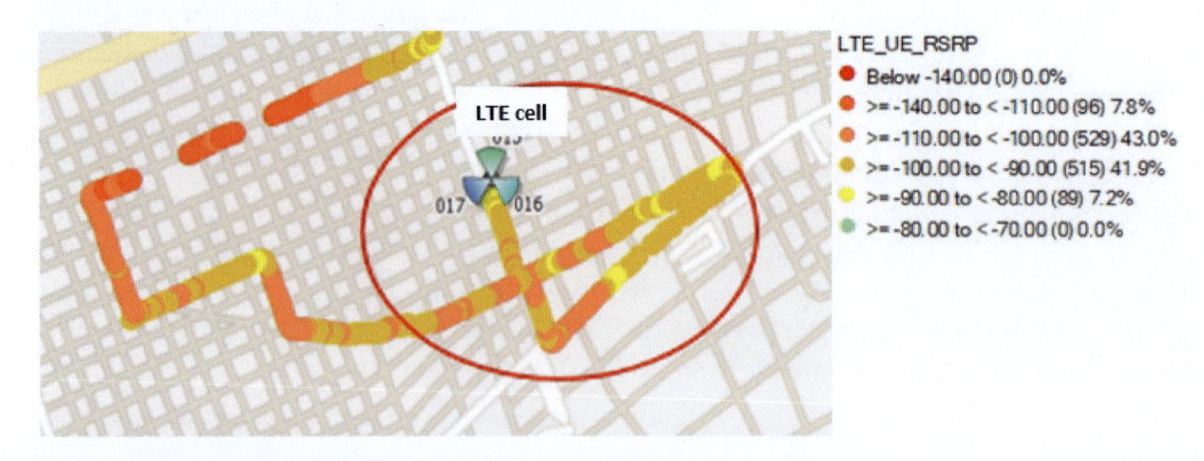

Figure 9.45 Weak LTE cell center coverage of high antenna height

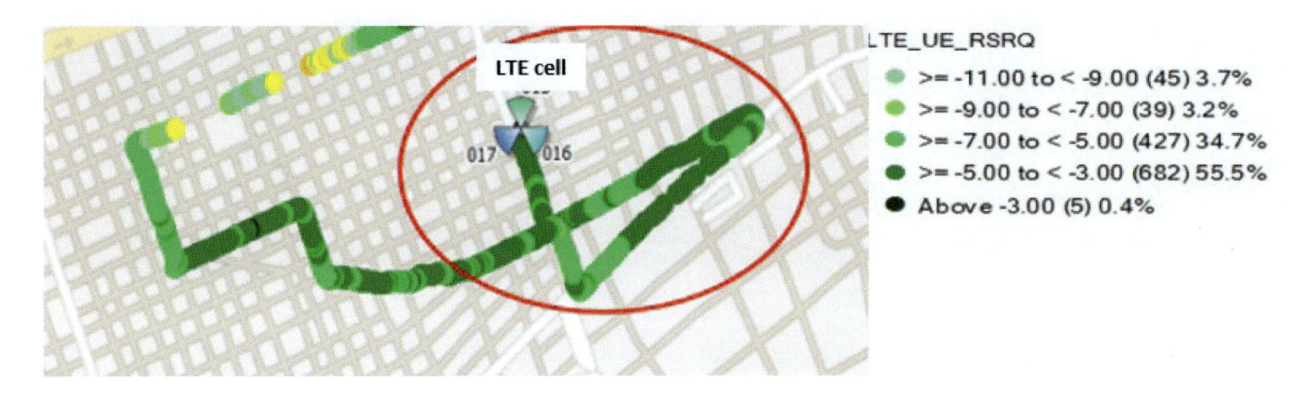

Figure 9.46 good quality at cell center of unloaded LTE cell

Another issue observed on this cell shown below was bad UL throughput for even a single user. The reason was found to be UL interference as high UL RSSI was recorded as shown in figure 9.47. The suggestion was to tune UL power control parameters and adjust UL/DL coverage imbalance by adjusting cell antenna tilt

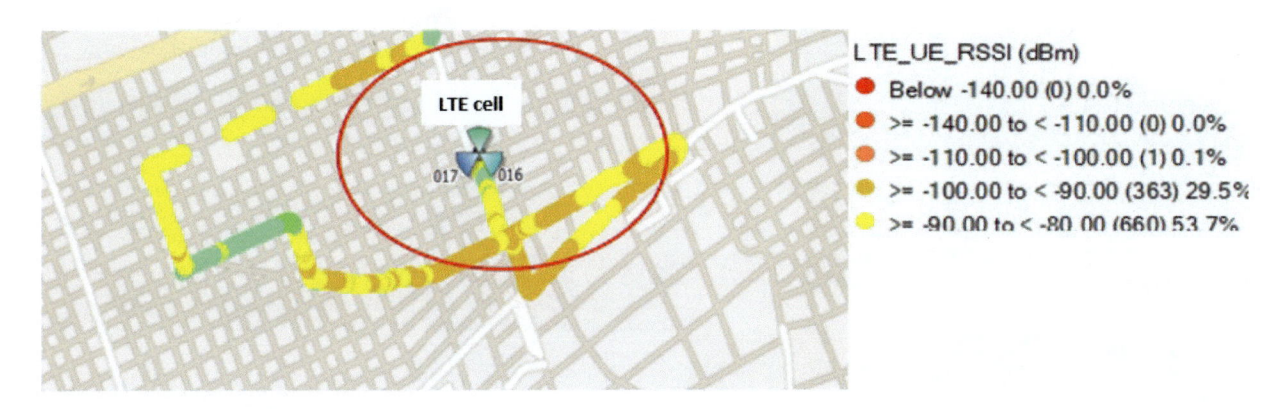

Figure 9.47 Case of LTE UL interference due to high UL RSSI

RF network optimization is vast subject and needs extensive learning and training. Operators who invest capacity building in this area will surely emerge as big competitive in the mobile telecom market. In the examples above cases, we presented snapshot from live network drive test. However, when doing drive test, reading L3 messages are also very important and interpreting their output.

Case 12: Some techniques of improving LTE coverage is illustrated below

Parameter settings
QrxLevMin,
PoNominalPUSCH/PUSCH,
PDCCH power boosting
RS power boosting
Cell selection / PRACH

far overshooting cell coverage
Causes UL synchronization loss

improve cell edge SINR
with ICIC

UL coverage
Weak ➔ high UE TX power
Detected with smaller PHR and UL SINR

Proper site settings
Tilt, azimuth, height,
location, VSWR

reduce UL interference
Detected with high UL RSSI, high UL BLER

lower frequency
Increases coverage

RS power boosting (pA, pB)

transmit diversity for
Improving cell edge coverage

	RSRP	RSRQ	UL SINR	RS SINR	PHR	UL RSSI	UE TX
DL coverage issue	< -115dBm	< -18dB interference		< 0dB			
UL coverage issue			< 0dB		< 4dB	High	High

We end our discussion on network optimization with further tips as shown in the boxes below

Interoperability optimization between LTE and UMTS cells

When we say interoperability between cellular systems, we mean cell reselection and handover. Cell reselection is idle mode behavior and can be done by the phone based on information in SIB messages. Handover is done by the network based on measurement report from the phone

When optimizing interoperability between LTE and UMTS cells we want the phone to always select high priority LTE cell

There are three things we can optimize in this case

- LTE to UMTS earlier than required reselection or handover while still LTE signal is good
- LTE to UMTS late reselection or handover when LTE cell has exceeded beyond satisfactory levels (RSRP < -120dBm)
- Ping pong or frequent reselection and handover between the cells

Cell reselection parameters	Handover event parameters
LTE (SIB1, SIB6)	LTE (A1, A2, B1)
UMTS (SIB11, SIB12, SIB19)	UMTS (2D, 2F 3A)
Reselection timer is the time between fulfillment of reselection criteria and execution	Time to trigger (TTT) is the time between event detection by phone and event triggering or reporting

Smaller TTT in handover configuration results in smaller delay and hence ping pong in area where there are many fluctuating signals, but increase handover success rate. If TTT parameter is increased, handover is delayed and hence ping pong is minimized. But if it is configured for macro cells, handover success rate will increase. Actual drive test KPI should be used to see the effect of modifying TTT

Different TTT configuration will be applicable for each cell based on coverage, capacity and user speed. Cell border should be optimally planned.

Similar argument applies for cell reselection timer. If reduced cell reselection will be delayed and increased otherwise

Latency in LTE cell is very high and users complain bad VOIP experience

Voice over IP (VOIP) is getting much appreciation in cellular networks than traditional voice calls. Virtually everyone with LTE capable phone and with internet uses VOIP calls using application like WhatsApp. Video calls over internet is also widely used

It is annoying when on VOIP call the voice cracks, delays and takes longer time to reconnect. Latency in LTE in one of the most important KPIs to be improved in the early phase of network construction

In drive test activity, ping will be used to test latency. Every part of the network such as air interface, backhaul transmission, core network, and data center can contribute to the overall latency and packet delay. Typically a ping latency of 50ms in loaded network is satisfactory

Some of the ways in which LTE latency could be improved are as follows

ping latency comparison between use at in the cell and test user connected to the core network
transmission latency should be reduced by optimizing signal CINR during installation and correct VLAN configuration
in RRC connected state, contention free random access can also reduce it
prescheduling in which the UE sending uplink request for PUSCH step is removed
improving RSRP and RSRS, reducing UL interference, good neighbor relationship plan, microcell and good PCI plan can also improve KPIs including latency

Introduction to 5G NR R15

5G is the fifth generation of 3GPP cellular standards. The radio access part is called new radio (NR). The first release in 2015 was R15 and was primarily called **non-standalone (NSA)** mode in the sense that it was designed to work with LTE EPC. So the LTE was providing both user plane and control plane, while the NR was intended to provide higher throughput

3GPP NR R15 and R16 main features are listed below

R15
- mMTC
- Service based architecture
- Slicing
- V2X
- API exposure
- WLAN and unlicensed spectrum

R16
- uRLLC
- V2X
- Integrated access and backhaul (IAB)
- eCAPIF
- Satellite access in 5G
- NR-U

To rollout R15, there are several options. One option for startups is to re-use the existing LTE EPC core and just add 5G NG-gNB to boost cell throughput. The UE will have simultaneous connection to both LTE and NR where both are co-sited. This is called EUTRAN-NR DUAL CONNECTIVITY (EN-DC). We will talk about this a little deeper in a moment.

5G core network was designed to add enhancements to the LTE EPC as explained below. In service based architecture, 5GCN functions can be deployed on the cloud. So instead of talking about nodes like MME, NE40, now the discussion will be what functions the operator needs to achieve its business target. Logical network slicing will allows different use cases to be achieved independently. One could configure eMBB, uRLLC, and mMTC slices

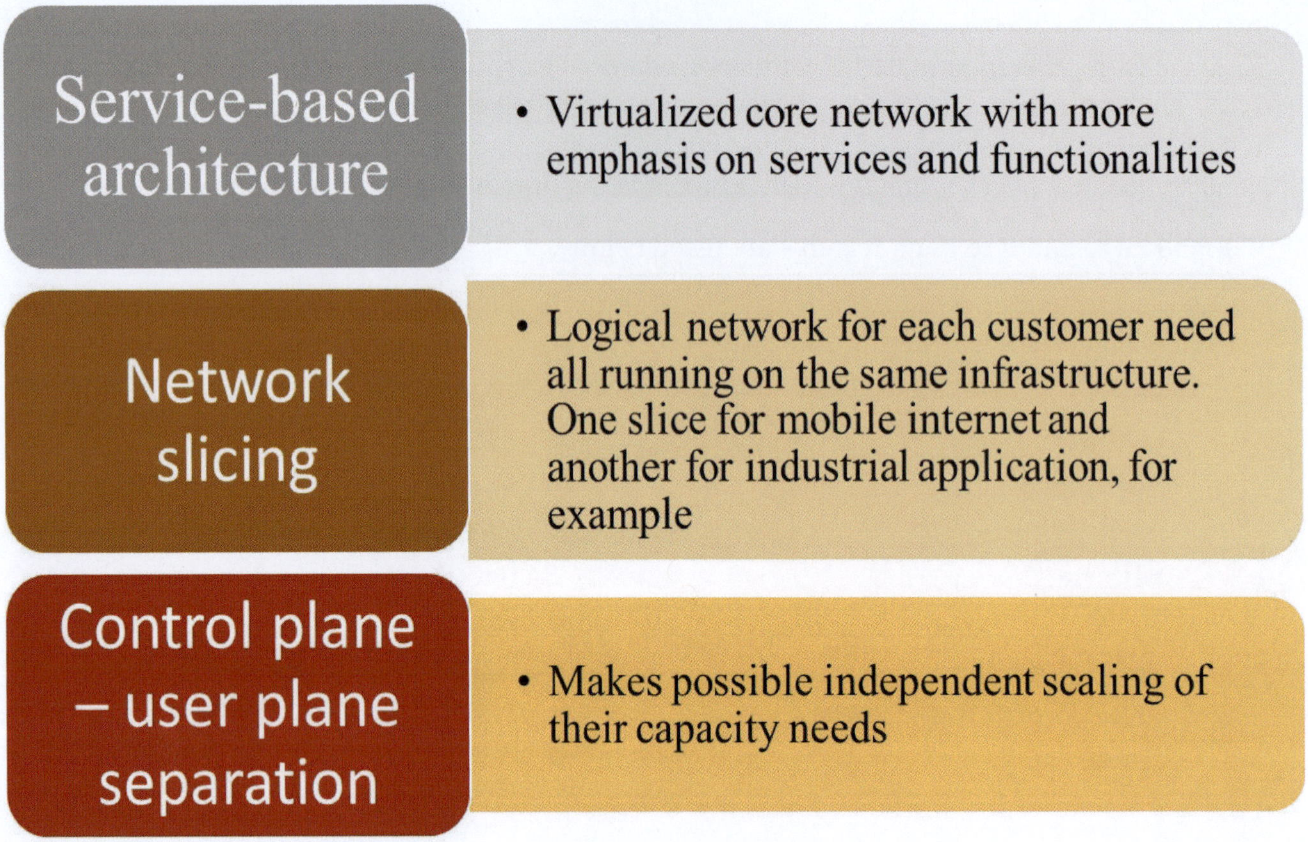

Service-based architecture
- Virtualized core network with more emphasis on services and functionalities

Network slicing
- Logical network for each customer need all running on the same infrastructure. One slice for mobile internet and another for industrial application, for example

Control plane – user plane separation
- Makes possible independent scaling of their capacity needs

Figure 9.48 5GCN new enhancements

From above, 5G core network (5GCN) focuses more on services and functionalities than nodes and devices. Thus shown below is service-based high level architecture of 5GCN

Cellular network planning fundamentals

Figure 9.49 service based 5GCN architecture

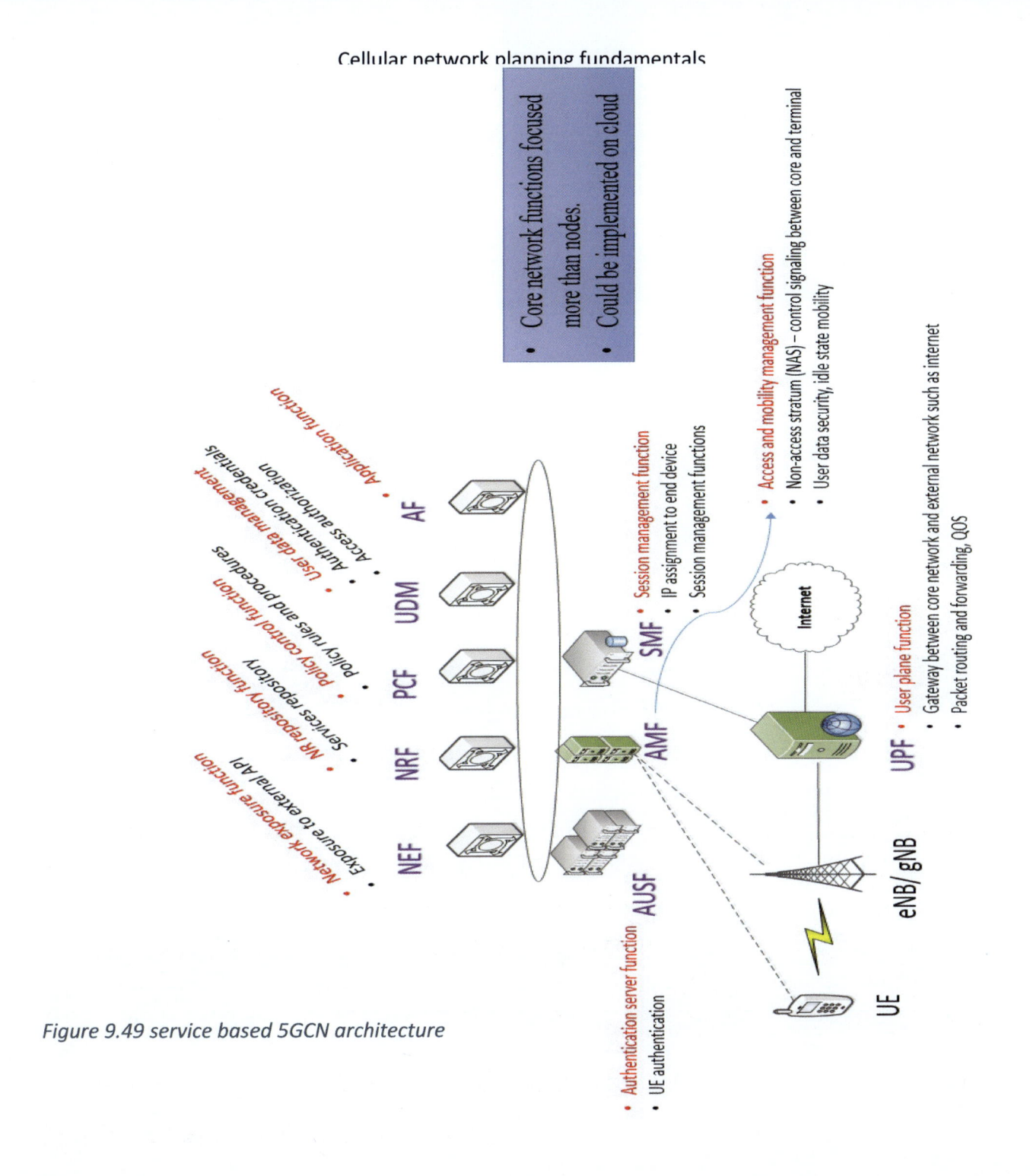

In **SA**, the NR is connected to 5GCN. Both user and control planes are handled by the 5GCN as shown in figure 9.50

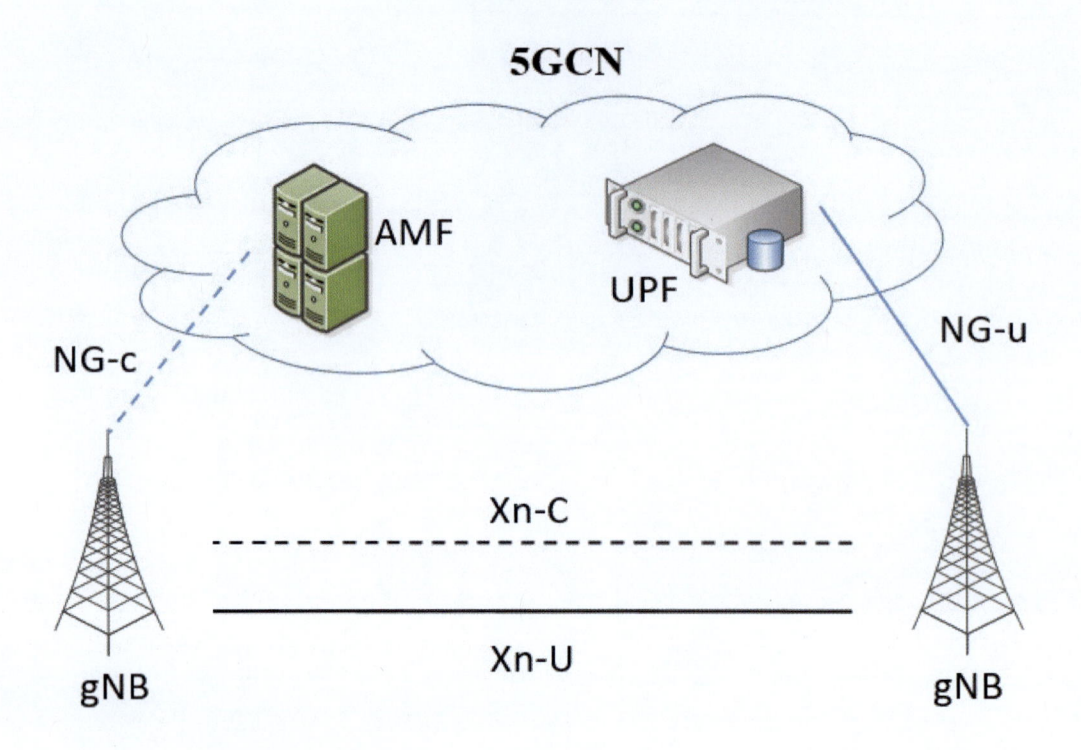

Figure 9.50 NR standalone architecture

Network slicing can be used to achieve the different use cases of the 5GCN. For example one slice could be for mobile broadband, another one for machine type communication and so on. The 5G use cases are shown in figure 9.51 below

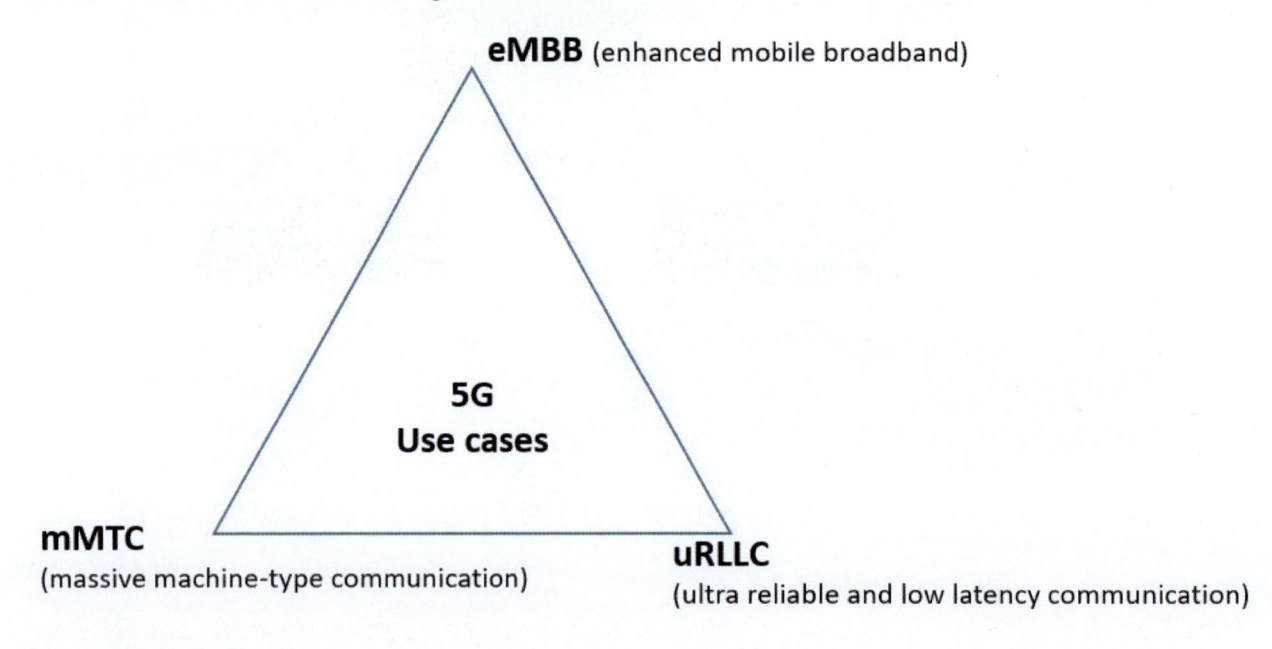

Figure 9.51 5G uses cases

We also need to talk about spectrum allocation for 5G NR. Two frequency bands in 3GPP R15

- FR1 includes all existing and new bands below 6GHz
- FR2 includes new bands in the range 24.25 – 52.6 GHz

NR support both FDD and TDD. Lower frequencies for FDD and higher frequencies for TDD
For eMBB 5G use case, very high data rate will consider higher bands 24, 60GHz

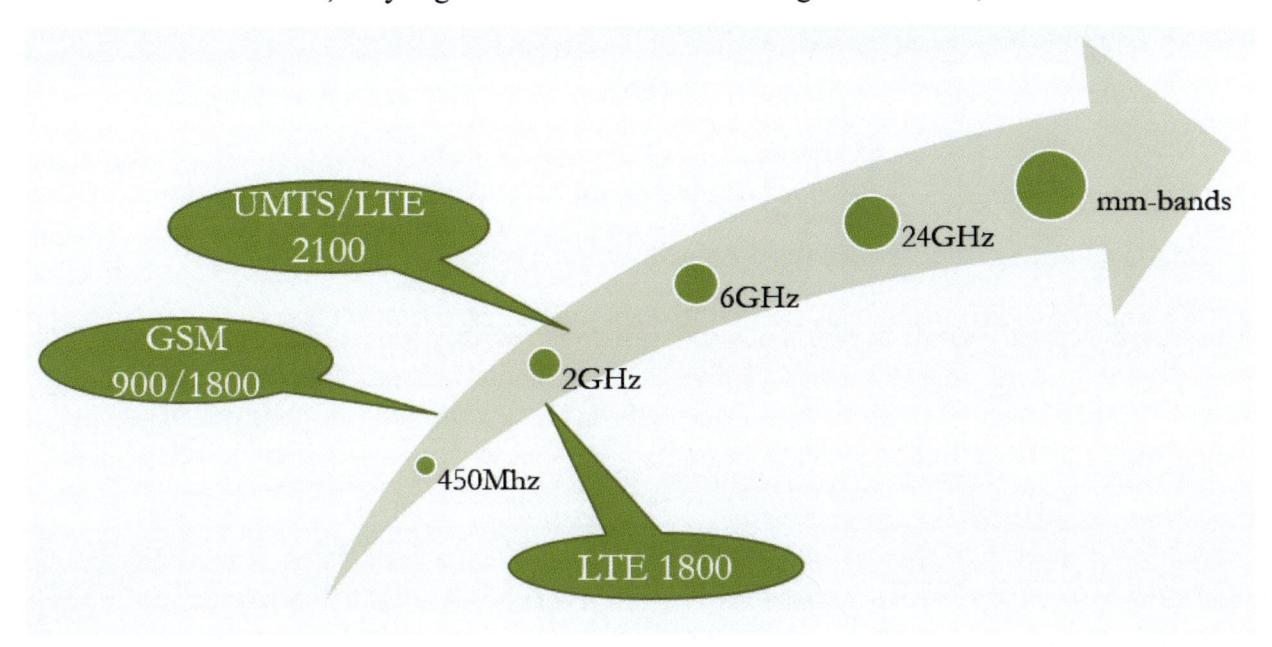

Figure 9.52 NR FR1 and FR2

TS 38 104 provides guidelines for radio transmission and reception such as NR-AFRCN and channel raster configurations

Table 5.4.2.1-1: NR-ARFCN parameters for the global frequency raster

Frequency range (MHz)	ΔF_{Global} (kHz)	$F_{REF-Offs}$ (MHz)	$N_{REF-Offs}$	Range of N_{REF}
0 – 3000	5	0	0	0 – 599999
3000 – 24250	15	3000	600000	600000 – 2016666
24250 – 100000	60	24250.08	2016667	2016667 – 3279165

The formula relating NR ARFCN and the corresponding center frequency is

$$F_{REF} = F_{ref\_offs} + \Delta F_{Global}(N_{REF} - N_{Ref_{offs}})$$

Where

$F_{REF} = center\ frequency$

$\Delta F_{Global} = channel\ raster$

$N_{REF} = NR\ ARFCN$

Let us take n7 (2600) in appendix B as an example which falls in the range 0 – 3000MHz

$F_{REF} = UL / DL = 2530 / 2650$

$2530 = 0 + 0.005(N_{REF}) = N_{REF} = 506{,}000 \quad for\ UL$

$2650 = 0 + 0.005(N_{REF}) = N_{REF} = 530{,}000 \quad for\ DL$

Hence center frequency UL / DL 2530 / 2650 corresponds to NR ARFCN 506,000 / 530,000

NR offers lower **latency** than LTE. Some of the NR design features to support low latency include

- Mini-slot transmission – fractions of slots useful in mm-wave
- MAC and RLC header structures designed for low latency
- RS and DL control signals located at the beginning of each transmission

In the first rollout of 5G, many operators deploy the non-standalone (NSA) mode of the NR. NSA will use the already existing LTE EPC. So both the LTE eNB and NR gNB will share the same spectrum as shown in figure 9.53. The LTE master cell will provide both control and user planes while 5G gNB will boost cell throughput in the user plane

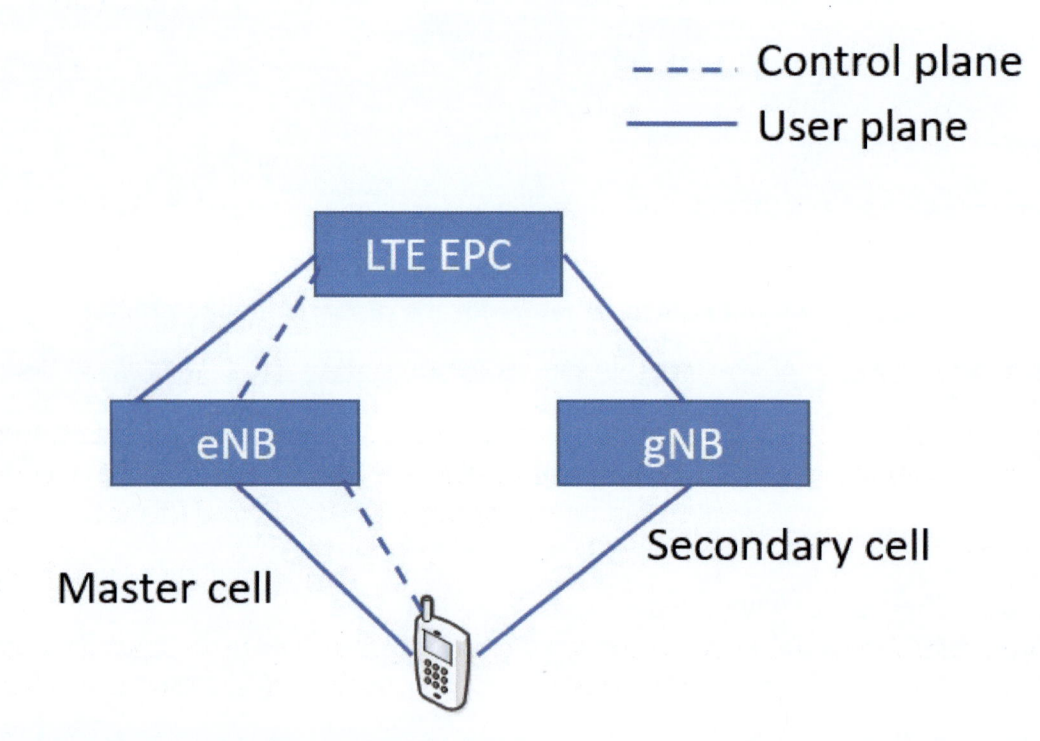

Figure 9.53 NR Non standalone architecture

Both LTE and NR co-existing in the same spectrum is shown in figure 9.54 below. Both share the spectrum resources in band 7 (n7) of FR1

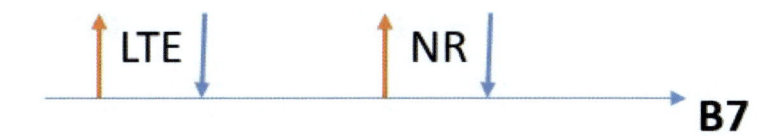

Figure 9.54 LTE NR spectrum co-existence

This allows **dual-connectivity** in which the UE will have simultaneous connection to both LTE and NR. The throughput is then aggregated. This means the UE is receiving two downlink transmission for eNB and gNB. However only one uplink transmission is allowed from the UE in a process called single-TX transmission coordinated by gNB and eNB schedulers

Single-TX transmission from the UE is allowed in the uplink by gNB and eNB schedulers coordination

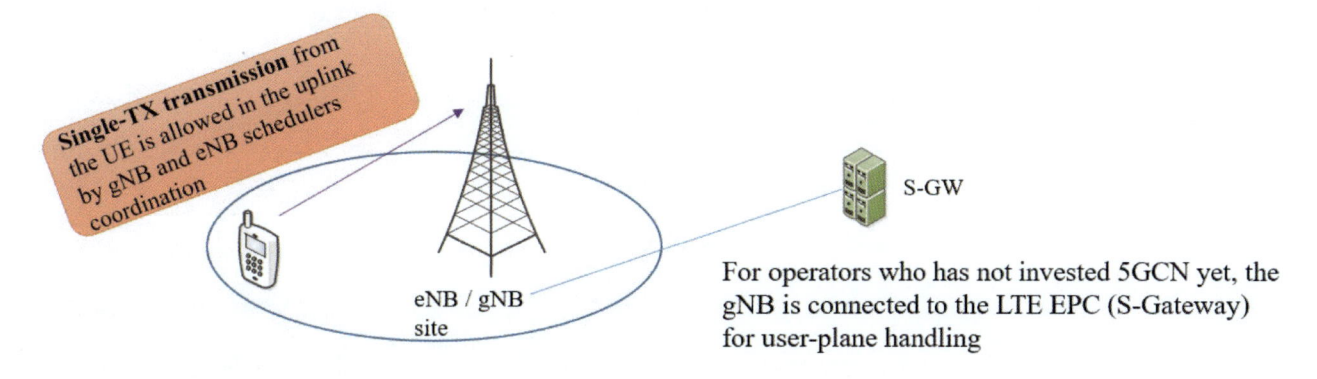

For operators who has not invested 5GCN yet, the gNB is connected to the LTE EPC (S-Gateway) for user-plane handling

Figure 9.55 single UL in EN-DC

The MN sends to the UE the *RRCConnectionReconfiguration* message including the NR RRC configuration message, without modifying it.

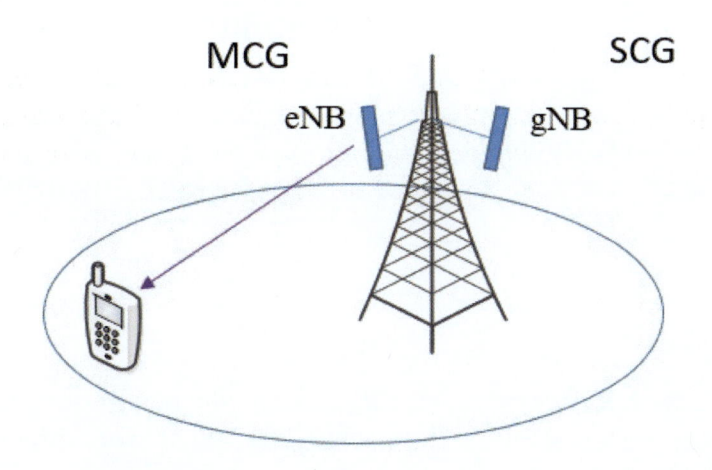

Figure 9.56 Initial connection to the MCG

Some of the parameters received from the EUTRAN to the UE for EN-DC are as follows:

To select the SCG, the EUTRAN shall send SIB2 to the UE

```
┌─────────────────────────┐
│   SIB2 (EUTRAN – UE)    │
│  PLMN-info-r15 = TRUE   │
└─────────────────────────┘
```

The UE shall also receive the EUTRAN UE capability enquire to ensure it supports EN-DC

```
UECapabilityEnquiry (EUTRAN – UE)

eutra-nr-only-r15 = TRUE
```

The UE shall also have information related to addition and release of SCG from RRC connection reconfiguration from EUTRAN

```
RRCConnectionReconfiguration-v1510 (EUTRAN – UE)

endc-ReleaseAndAdd-r15 = TRUE
SCellGroupToAddModList-r15 = 1
sCellState-r15 = ACTIVATED
```

The figure below summarizes messages during addition of SCG (TS 37.340)

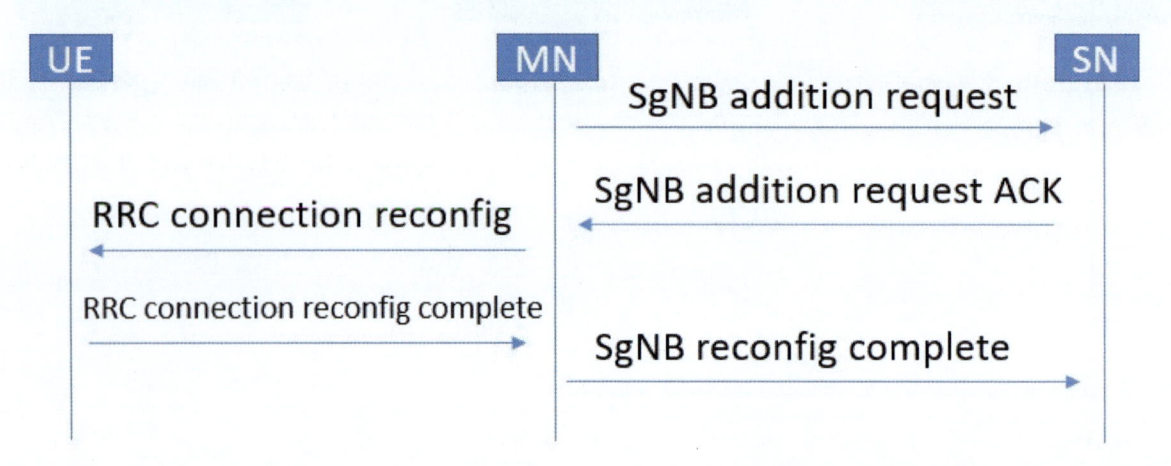

Figure 9.57 EN-DC SCG addition

In other scenario where the NR is deployed in FR2, a supplementary uplink (SUL) is configured in FR1 to compensate for the larger pathloss associated in FR2. This is shown in figure 9.58 below

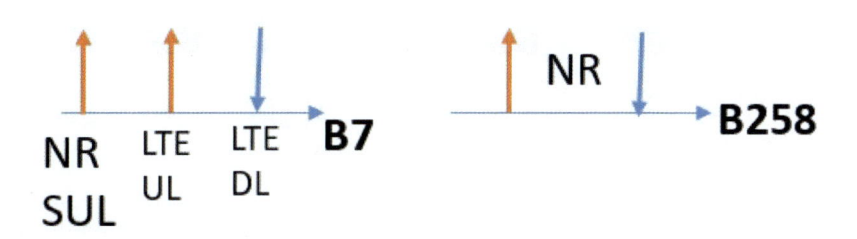

Figure 9.58 SUL (supplementary uplink)

NR has flexible frame structure. You remember from figure 9.16 where we discussed LTE frame structure. We said an LTE radio frame is 10ms in the time domain and 12 subcarriers in the frequency domain. Each subcarrier in LTE is 15 kHz of bandwidth. This is also true for NR. Thus LTE and NR can coexist in the 15 kHz frame structure.

Besides 15 kHz, NR has also got other frame structure numerologies. Flexible subcarrier spacing (**SCS**) 30 kHz, 60 kHz, 120 kHz, and 240 kHz

For example, the 15 kHz frame has 14 OFDM symbols each of 1ms duration, and the 30 kHz frame has 14 OFDM symbols each of 0.5ms duration.

Resource element (RE) = 1 OFDM symbol by 1 subcarrier

In LTE resource block (RB) is 14 OFDM symbols by 12 subcarriers

In NR resource block (RB) is only defined for frequency domain because time domain allows mini-slot transmission for low latency support

LTE supports 100 resource blocks at 20 MHz while NR supports 275 resource blocks at 400 MHz

For large SCS, the symbol duration will be smaller. This would beneficial in uRLLC use case to support low latency. But large SCS leads to smaller cyclic prefix (CP). Hence mini slot transmission is adopted to support low latency communication.

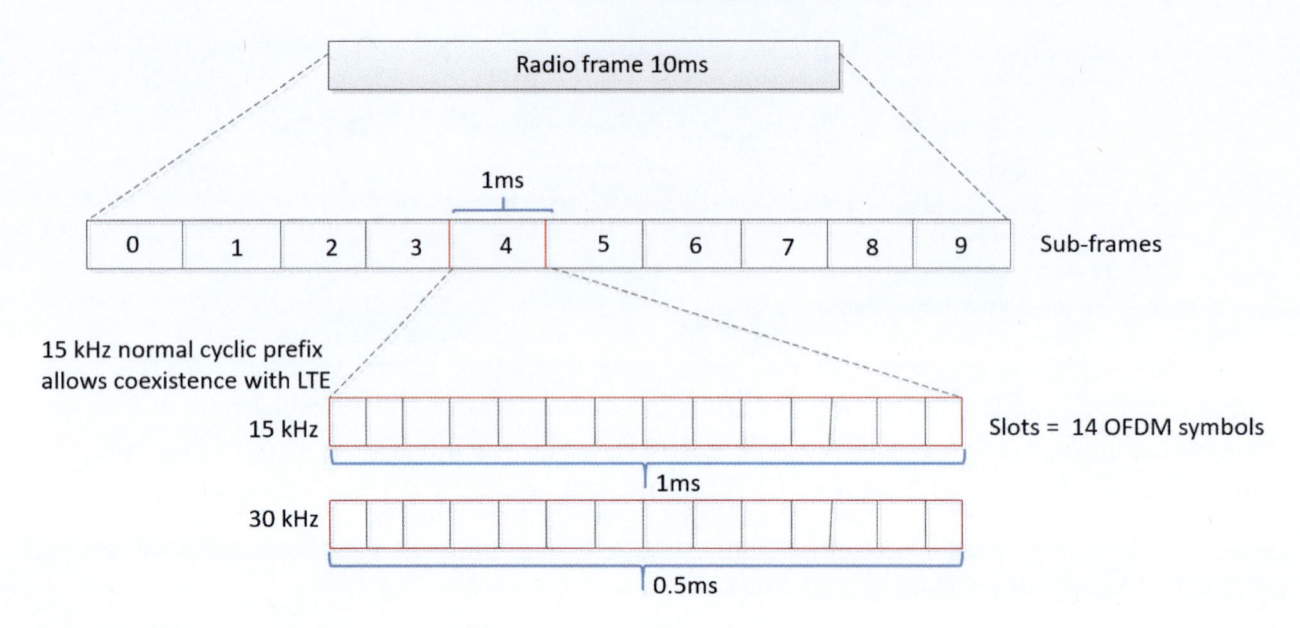

Figure 9.59 NR frame, subframe, and slot

The NR radio network access performs several functions. Some are listed below

| Scheduling | Radio resource management | Coding and re-transmission protocols |

The UE and the NR gNB need protocols in order to understand each other and communicate. The figure below summarized the NR air interfaces protocol stack (the SDAP protocol is required in the standalone mode only for quality of service (QOS) handling

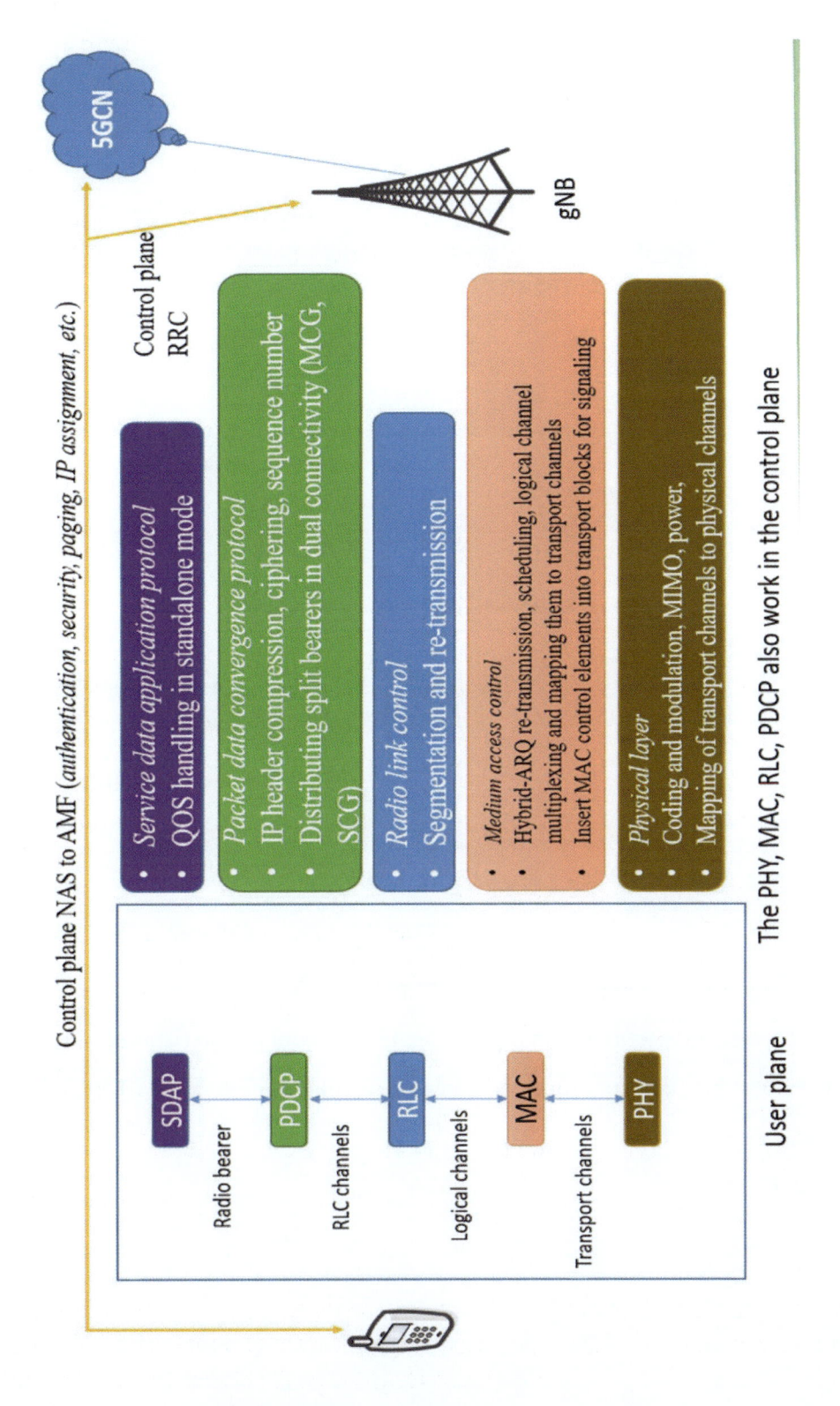

Figure 9.60 NR air interface protocol stack

The MAC layer multiplexes logical channels from RLC to transport channels which is then is mapped into physical channels by the physical layer

Logical channel is characterized by information type they carry

- When they carry user data → traffic channel
- When they carry signaling data → control channel

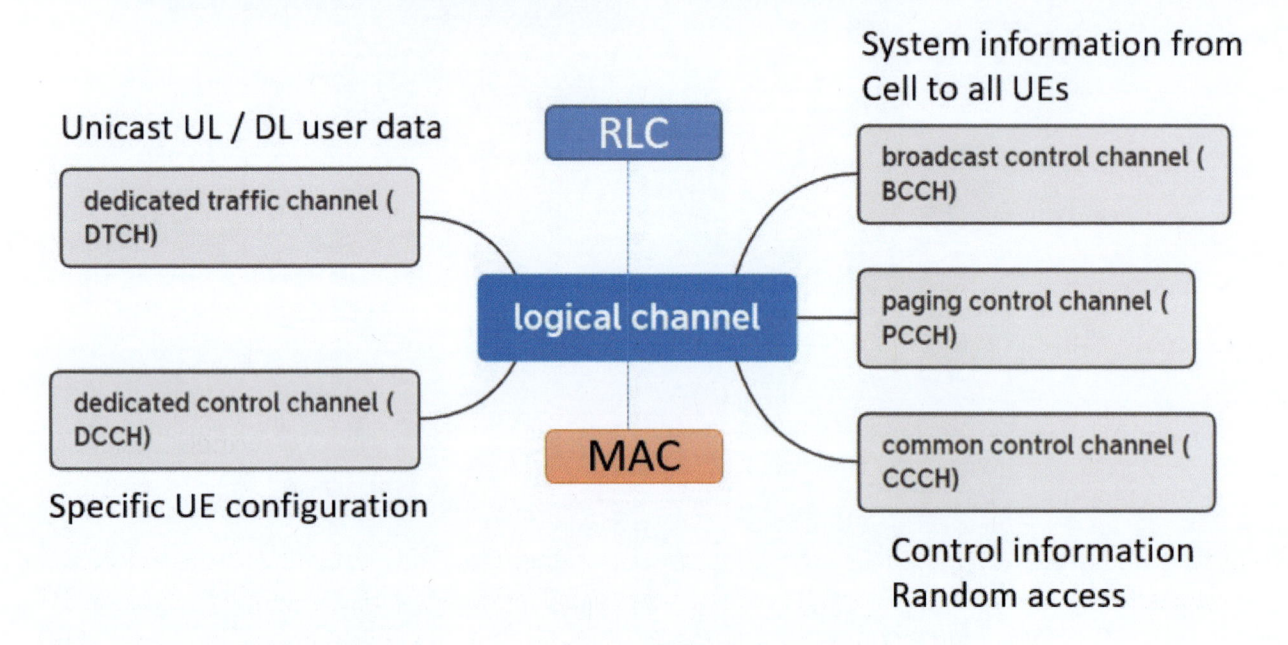

Figure 9.61 NR logical channels

Transport channel defines the way data is transmitted over the radio channel
- Thus transport channel organizes data into transport blocks
- Each TTI, one transport block is sent to the UE (except 4-layer spatial multiplexing)
- Each transport block is specified by transport format (TF) such as ModCod, size, antenna mapping
- Different TFs → different data rate by MAC layer

NR transport channels

Uplink data
- uplink shared channel (UL-SCH)
- random access channel (RACH) — Does not carry transport blocks

transport channel (MAC / PHY)

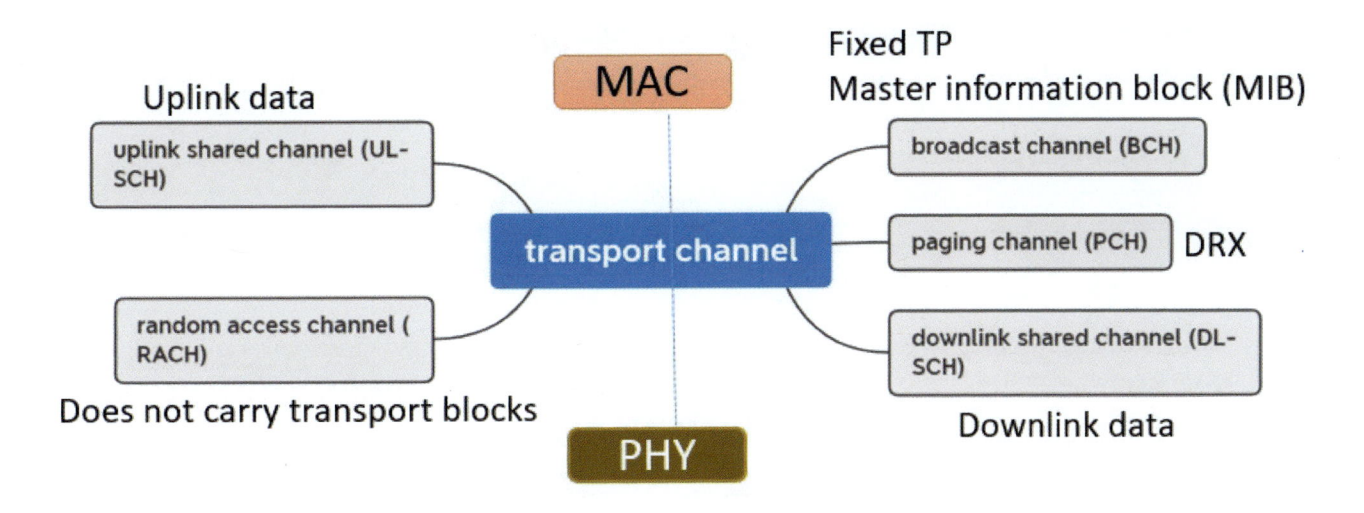

Downlink data
- broadcast channel (BCH) — Fixed TP, Master information block (MIB)
- paging channel (PCH) — DRX
- downlink shared channel (DL-SCH)

Figure 9.62 NR transport channels

Physical channel is time-frequency resource used for transmission of transport channel

- Each transport channel is mapped to corresponding physical channel (for example DL-SCH → PDSCH), except L1/L2 control channel that carry DCI / UCI information that do not have transport channels (PDCCH/PUCCH)

physical channel

- physical broadcast channel (PBCH) — MIB and SIBs for accessing the network; Activated DL bandwidth parts
- physical uplink shared channel (PUSCH) — UL data
- physical uplink control channel (PUCCH)
  - Request resource for UL data
  - Hybrid-ARQ acknowledgement
  - Channel state (CSI) report
- physical downlink shared channel (PDSCH) — Mainly for unicast DL data. Also RA response, SIBs
- physical random access channel (PRACH) — Random access
- physical downlink control channel (PDCCH)
  - Scheduling decision for PDSCH decoding
  - Scheduling grands for PUSCH

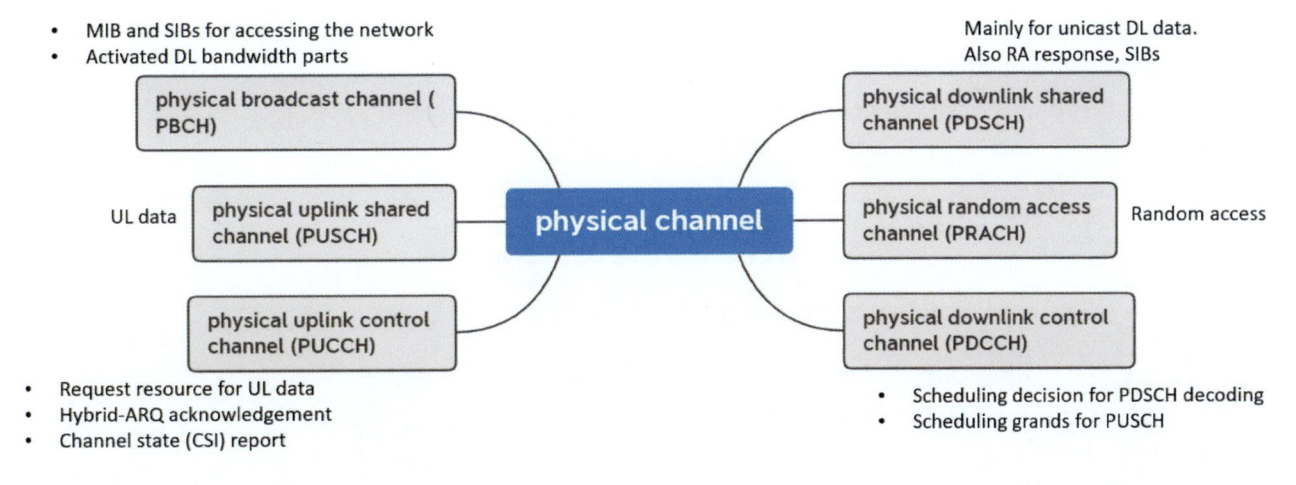

Figure 9.63 NR physical channels

To achieve network slicing in NR, quality of services (**QOS**) is configured in the RAN. In LTE, the PCRF dynamically generates QOS policies in which the P-GW implements. The eNB sets bearers to UEs and schedules data based on QOS parameters.

In 5GCN SA, the UPF handles QOS service handling and maps different QOS requirements (latency, data rate, etc.) into QOS flows. The gNB maps the QOS flows into data radio bearers (DRB). The UE learns the mapping via RRC messages in which it is explicitly configured. On the air interface, the SDAP protocol will handle the QOS. The QOS mappings is illustrated in figure 9.64 below

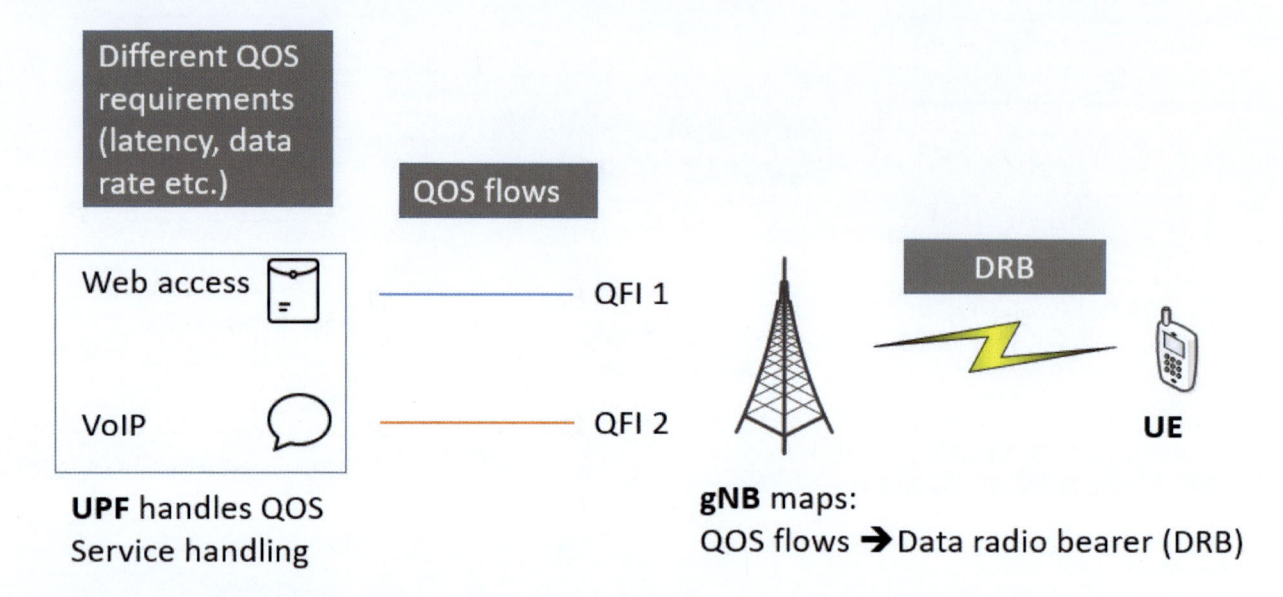

Figure 9.64 NR QOS handling

Another concept in NR is **bandwidth part.**

Bandwidth part is a subset of a contiguous common resource blocks. Initial bandwidth part in the UL and DL is defined. UE will receive DL bandwidth part from physical broadcast channel (PBCH) and UL bandwidth part from physical downlink control channel (PDCCH)

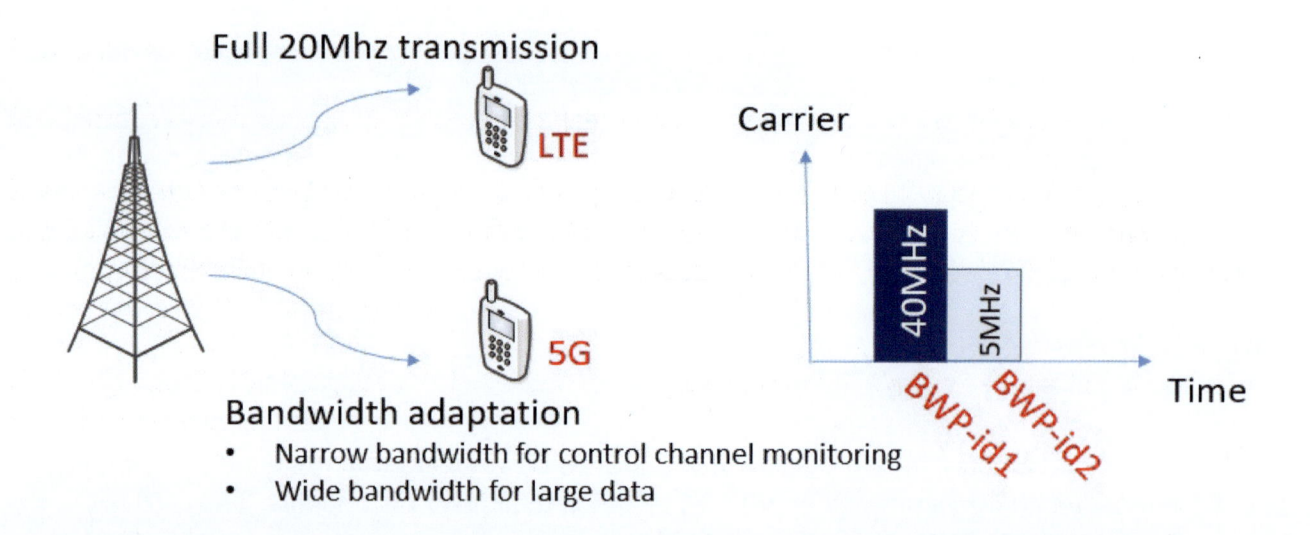

Figure 9.65 NR bandwidth parts

TS 38.331 Sec 6.3.2 provides the various parameters to be set to active bandwidth parts

Like LTE the network needs channel quality report from the UE to adjust data scheduling. UL and DL **channel sounding** signals help the UE to send such report

Downlink channel sounding

LTE R8	*Cell specific reference signal* (C-RS) • Transmitted over the entire 20 MHz in every TTI (1ms sub-frame) • Always present over the entire cell
LTE R10	CSI-RS (*channel state information reference signal*) introduced Explicit configuration on the device
NR	CSI-RS used for DL channel estimation • Not always on • Configured for a give DL *bandwidth part*

Uplink channel sounding

Sounding reference signal (SRS) is used for uplink channel estimation

Scheduling is the assignment of uplink and downlink resources to the devices in the cell

- Frequency domain → resource block, time domain → OFDM symbol
- Time – frequency resources are dynamically shared between users (*dynamic scheduling*)

To estimate DL channel quality for downlink scheduling

- UE transmits channel state information (CSI)

To estimate UL channel quality for uplink scheduling

- UE sends sounding reference signal (SRS)
- Buffer status report (BSR), power headroom (PHR)

DL scheduling
- Resource blocks to transmit PDSCH
- Transport format (transport block size, modulation scheme)
- gNB control logical channel multiplexing

UL scheduling
- Resource for PUSCH
- gNB controls transport format
- UE decides logical channel multiplexing

Figure 9.66 data scheduling

Now let us assume the UE is in NR coverage and wants to perform initial access after powering on. The initial access steps look pretty similar to LTE with some changes

Cell search is performed by the UE when

- It enters coverage of the network
- In idle mode for cell reselection and connected mode for handover
- Each NR cell periodically transmits synchronization signal block (SsBlock) in the downlink
- SsBlock = PSS + SSS + PBCH

After cell is found and camped on, devices in RRC idle/in-active perform *random access*

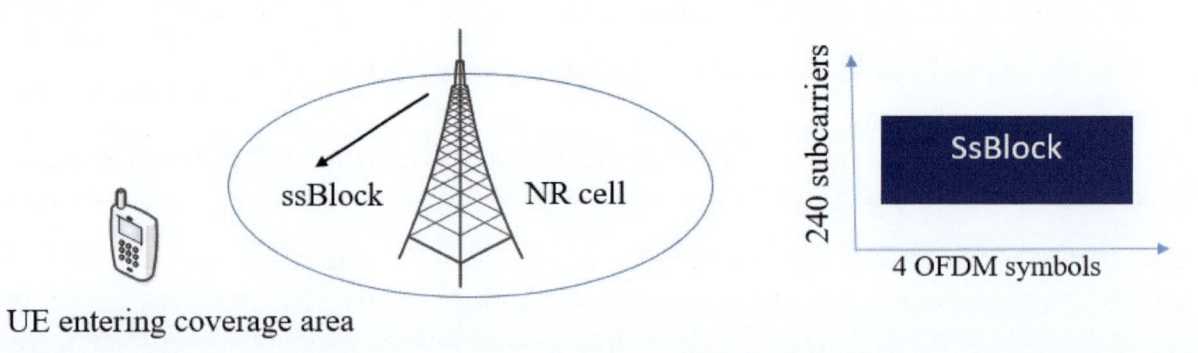

UE entering coverage area

Figure 9.67 NR ssBlock

The over NR initial access procedure is summarized below. In LTE the PCI ranges from 0 – 503 while in NR the PCI ranges from 0 – 1007

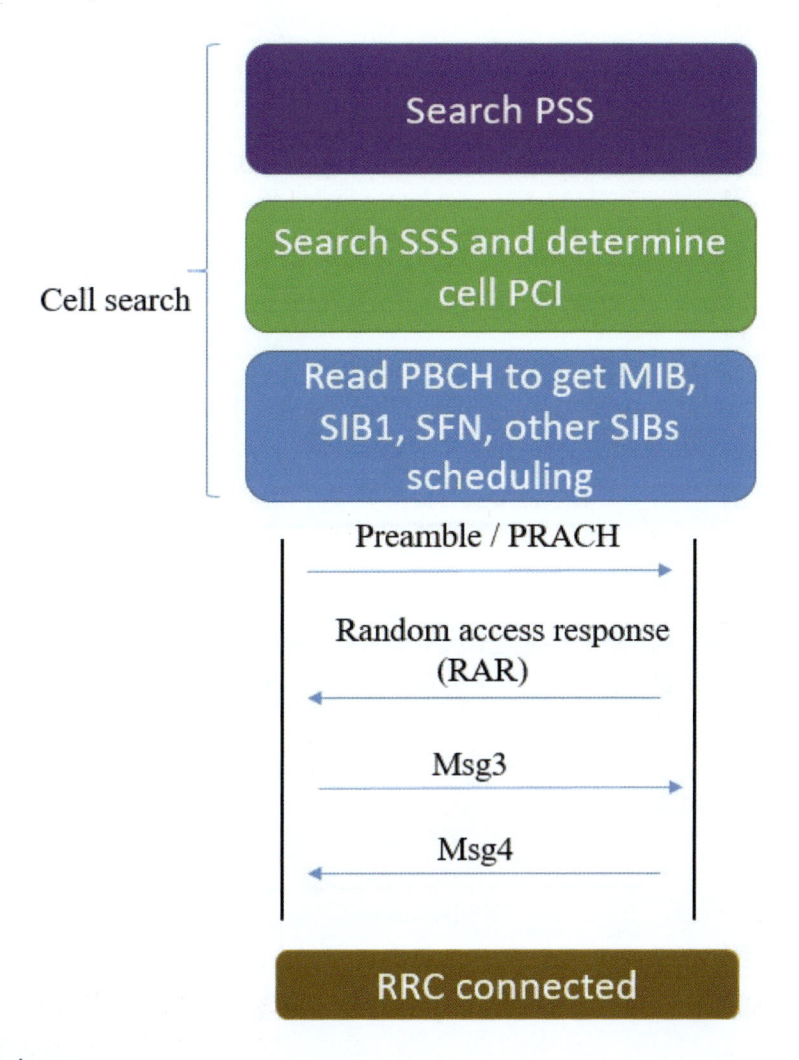

Figure 9.68 NR initial access

Some of the differences between NR and LTE initial access is summarized in the table below

LTE	NR
PSS/SSS/PBCH located at center of the carrier, searched at each position of *carrier raster*	Faster cell search by only locating limited set of *synchronization raster*
5ms PSS/SSS/PBCH periodicity	20ms SsBlock periodicity
All SIBs periodically broadcast in the cell	SIB1 periodically broadcast to perform random access. Other SIBs sent upon request by connected device

Multiple antenna transmission (**MIMO**) is used in NR like LTE to combat channel fading at cell edge users and increase throughput for cell center users. In NR massive MIMO will be more important in FR2

Beamforming is directing the transmit power from gNB to a certain UE in order to increase throughput in that direction

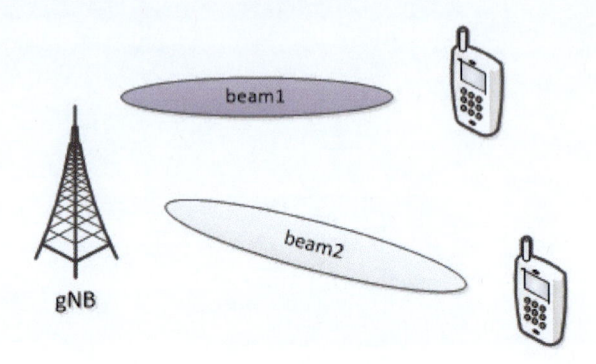

Figure 9.69 NR beamforming

Because of UE mobility and beam blockage of obstruction, each UE will have a **beam pair** (transmit beam and corresponding receive beam). The beam pair can be direct or reflected by obstructions. Figure 9.70 below shows downlink beam pair

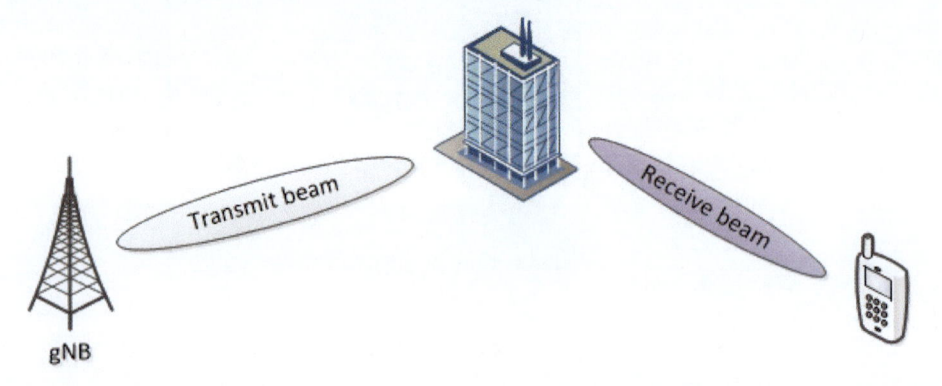

Figure 9.70 beam pair

The above downlink beam pair can be used in the uplink as well and is called *beam correspondence*

When UE is making connection with the cell, an initial beam establishment is performed. In this way the UE will receive different downlink ssBlock signals and their corresponding downlink beam pairs. The UE will then use the established downlink beam pair in the uplink as well

The UE will then select one ssBlock beam pair based on the configured ss-RSRP threshold

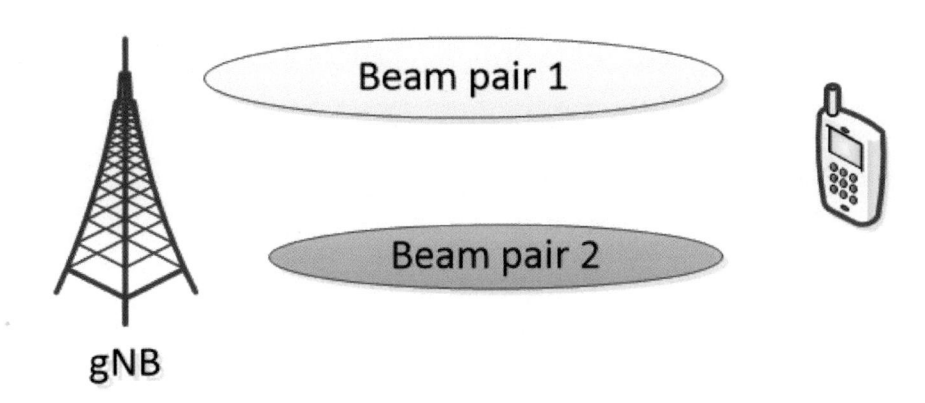

Figure 9.71 candidate downlink beam pairs

Now after an initial beam pair is established, beam adjustment is evaluated continuously due to user mobility and beam obstructions

In some cases, the established beam pair can fail. *Beam recovery* procedure is then initiated.

Beam pair can fail when (1) measured ssBlock < configured threshold (2) PDCCH BLER < configured threshold. Once this thresholds are exceeded and the beam pair failure is detected, the UE will then try to find a new beam pair. The UE will already have a configured a set of downlink ssBlocks and their corresponding beam pairs, and will identify the best among the set of the candidate SSB beams

```
BeamFailureRecoveryConfig: =
{    Rsrp-ThresholdSSB = 67
     CandidateBeamSSBList = 2      }        → TS 38 331
```

The UE will then send *beam recovery request* to the gNB after candidate beam identification

In this message the UE will attempt contention free Random Access to recover from beam failure. The following example parameters is used to configure the UE with CFRA RACH resources and candidate beams for beam failure recovery

```
BeamFailureRecoveryConfig: =
{    RootSequenceIndex-BFR = 0
     BeamFailureRecoveryTimer = ms40
     Msg1-SubcarrierSpacing = 30
                                   }        → TS 38 331
```

We end our discussion on 5G introduction with comparison between LTE and NR.

	LTE	NR
Carrier spacing	15 kHz	Flexible (15, 30, 60, 120, 240)
Cyclic prefix duration	4.7μs	Flexible (4.7, 2.3, 1.2, 0.59, 0.29)
Duplex	Two different frame structures for FDD and TDD	One common frame structure for FDD and TDD
Channel coding	Turbo	LDPC
Latency	larger – MAC and RLC protocols know the amount of data to transmit before any processing	Small – MAC and RLC allow processing without knowing amount of data to transmit
PDCCH	Transmitted using full bandwidth	Configured to occupy only part of bandwidth. Beamforming on PDCCH
Ultra-lean design	All signals on	ssBlock the only signal always on
Time-domain	10ms radio frame divided into 10 sub-frames each of 1ms duration @15 kHz	Same as LTE @15 kHz. Different time-domain @30, 60, 120, 240 kHz numerologies

Appendix A: Cellular FDD RF channels

GSM = ARFCN

UMTS = UARFCN

LTE = EARFCN

NR = NR-ARFCN

Band	UL FREQ	UL ARFCN	DL FREQ	DL ARFCN	Raster	BW
GSM-900	880 - 915	0 – 124 975 - 1023	925 - 960	0 – 124 975 - 1023	200kHz	35MHz
UMTS-2100	1920 – 1980	9612 – 9888	2110 - 2170	10592 - 10838	200kHz	60MHz
LTE-1800	1710 – 1785	19200 - 19949	1805 - 1880	1200 - 1949	100kHz	75MHz
UMTS-900	880 - 915	2712 - 2863	925 - 960	2937 - 3088	200kHz	35MHz
GSM-1800	1710 – 1785	512 - 885	1805 - 1880	512 - 885	200kHz	75MHz
LTE-800	832 – 862	24150 - 24449	791 - 821	6150 - 6449	100kHz	30MHz

Appendix A: Cellular FDD RF channels

LTE-2100	1920 – 1980	18000 – 18599	2110 - 2170	0 - 599	100kHz	60MHz
LTE-2600	2500 – 2570	20750 – 21449	2620 - 2690	2750 - 3449	100kHz	70MHz
NR – 2600	2500 – 2570	500000 – 514000	2620 – 2690	524000 – 538000	100kHz	70MHz

Appendix B: NR FDD FR1 bands (138 104 v15.4.0 table 5.2-1)

NR operating band	Uplink (UL) operating band BS receive / UE transmit $F_{UL,low} - F_{UL,high}$	Downlink (DL) operating band BS transmit / UE receive $F_{DL,low} - F_{DL,high}$	Duplex Mode
n1	1920 MHz – 1980 MHz	2110 MHz – 2170 MHz	FDD
n2	1850 MHz – 1910 MHz	1930 MHz – 1990 MHz	FDD
n3	1710 MHz – 1785 MHz	1805 MHz – 1880 MHz	FDD
n5	824 MHz – 849 MHz	869 MHz – 894 MHz	FDD
n7	2500 MHz – 2570 MHz	2620 MHz – 2690 MHz	FDD
n8	880 MHz – 915 MHz	925 MHz – 960 MHz	FDD
n12	699 MHz – 716 MHz	729 MHz – 746 MHz	FDD
n20	832 MHz – 862 MHz	791 MHz – 821 MHz	FDD
n25	1850 MHz – 1915 MHz	1930 MHz – 1995 MHz	FDD
n28	703 MHz – 748 MHz	758 MHz – 803 MHz	FDD
n34	2010 MHz – 2025 MHz	2010 MHz – 2025 MHz	TDD
n38	2570 MHz – 2620 MHz	2570 MHz – 2620 MHz	TDD
n39	1880 MHz – 1920 MHz	1880 MHz – 1920 MHz	TDD
n40	2300 MHz – 2400 MHz	2300 MHz – 2400 MHz	TDD
n41	2496 MHz – 2690 MHz	2496 MHz – 2690 MHz	TDD
n50	1432 MHz – 1517 MHz	1432 MHz – 1517 MHz	TDD
n51	1427 MHz – 1432 MHz	1427 MHz – 1432 MHz	TDD
n65	1920 MHz – 2010 MHz	2110 MHz – 2200 MHz	FDD
n66	1710 MHz – 1780 MHz	2110 MHz – 2200 MHz	FDD
n70	1695 MHz – 1710 MHz	1995 MHz – 2020 MHz	FDD
n71	663 MHz – 698 MHz	617 MHz – 652 MHz	FDD
n74	1427 MHz – 1470 MHz	1475 MHz – 1518 MHz	FDD
n75	N/A	1432 MHz – 1517 MHz	SDL
n76	N/A	1427 MHz – 1432 MHz	SDL
n77	3300 MHz – 4200 MHz	3300 MHz – 4200 MHz	TDD
n78	3300 MHz – 3800 MHz	3300 MHz – 3800 MHz	TDD
n79	4400 MHz – 5000 MHz	4400 MHz – 5000 MHz	TDD
n80	1710 MHz – 1785 MHz	N/A	SUL
n81	880 MHz – 915 MHz	N/A	SUL
n82	832 MHz – 862 MHz	N/A	SUL
n83	703 MHz – 748 MHz	N/A	SUL
n84	1920 MHz – 1980 MHz	N/A	SUL
n86	1710 MHz – 1780 MHz	N/A	SUL

Appendix C: NR TDD FR1 bands (138 104 v15.4.0 table 5.2-2)

NR operating band	Uplink (UL) and Downlink (DL) operating band BS transmit/receive UE transmit/receive $F_{UL,low} - F_{UL,high}$ $F_{DL,low} - F_{DL,high}$	Duplex Mode
n257	26500 MHz – 29500 MHz	TDD
n258	24250 MHz – 27500 MHz	TDD
n260	37000 MHz – 40000 MHz	TDD
n261	27500 MHz – 28350 MHz	TDD

Appendix D: NR transmission bandwidth configuration (TS 138 104 V15.4.0)

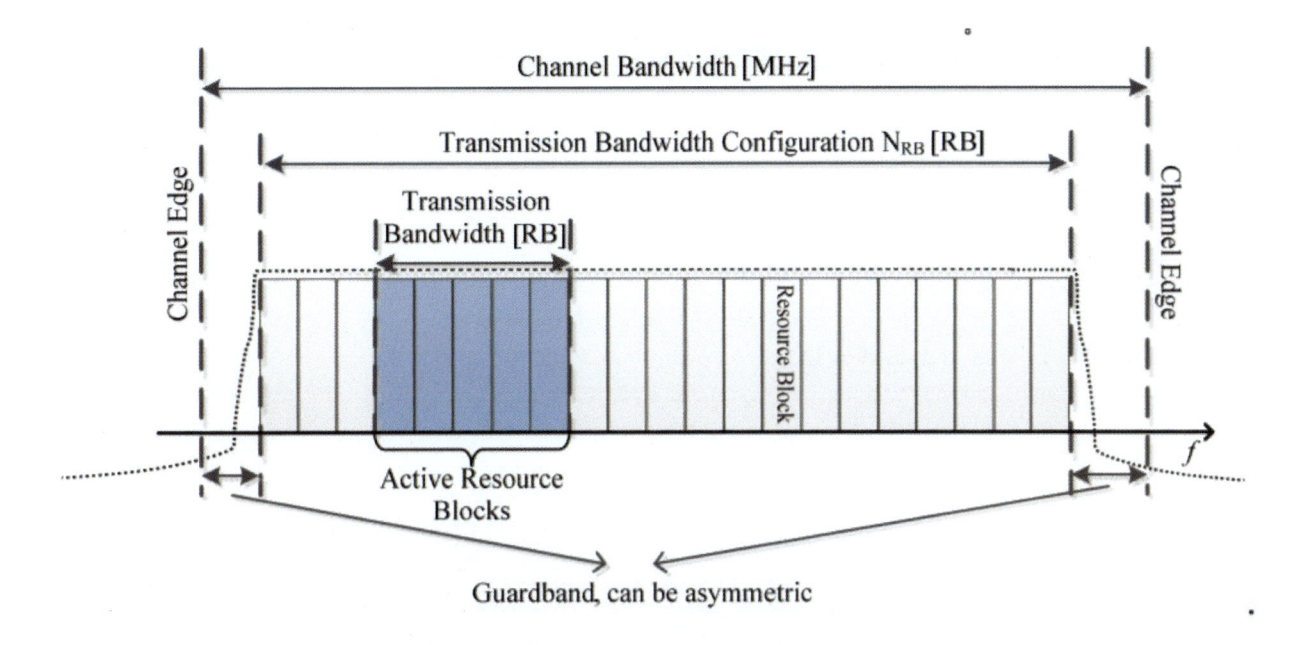

Table 5.3.2-1: Transmission bandwidth configuration N_{RB} for FR1

SCS (kHz)	5 MHz N_{RB}	10 MHz N_{RB}	15 MHz N_{RB}	20 MHz N_{RB}	25 MHz N_{RB}	30 MHz N_{RB}	40 MHz N_{RB}	50 MHz N_{RB}	60 MHz N_{RB}	70 MHz N_{RB}	80 MHz N_{RB}	90 MHz N_{RB}	100 MHz N_{RB}
15	25	52	79	106	133	160	216	270	N.A	N.A	N.A	N.A	N.A
30	11	24	38	51	65	78	106	133	162	189	217	245	273
60	N.A	11	18	24	31	38	51	65	79	93	107	121	135

Table 5.3.2-2: Transmission bandwidth configuration N_{RB} for FR2

SCS (kHz)	50 MHz N_{RB}	100 MHz N_{RB}	200 MHz N_{RB}	400 MHz N_{RB}
60	66	132	264	N.A
120	32	66	132	264

Appendix E: EN-DC operating bands

When deploying EN-DC in which eNB is MCG and gNB is SCG, the SCG can be used in FR1 or FR2.

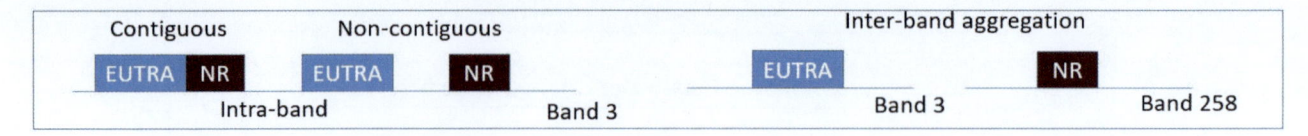

Intra-band non-contiguous EN-DC
```
DC_3_n3
Single UL allowed
```

Inter-band EN-DC including FR2
```
DC_3_n258
DC_7_n258
Single UL not allowed
```

Inter-band EN-DC within FR1
```
DC_1_n78
DC_3_n78
DC_7_n78
Single UL not allowed
```

For example, in intra-band non-contiguous EN-DC, the maximum aggregated bandwidth is 50MHz

Channel BW LTE	Channel BW NR	Maximum aggregated BW
5, 10, 15, 20	5, 10, 15, 20, 25, 30	50

Appendix F: Microwave channel arrangement

Microwave radio channels can be calculated from ITU-R recommendation F-series given in the table below. Use the latest annex.

Band	Recommendation
7	F.385
8	F.386
11	F.387
13	F.497
15	F.636
18	F.595
23	F.637
26	F.748
38	F.749

Let us take an example to calculate 7GHz of channel bandwidth 28MHz and TR spacing 245MHz using F.385-10

The spectrum licensed by the operator is 7425 – 7900MHz

With center frequency denoted as f_0, center frequency of low sub-band channel f_n, center of the high sub-band channel f'_n, then the frequencies of the cannel plan of the low sub-band is calculated from the following equations

$$f_n = f_n - 248.5 + 28n$$

The corresponding high sub-band channel arrangement is given by

$$f'_n = f_0 - 3.5 + 28n$$

Now the center frequency of the spectrum block is (7425 + 7900) / 2 = 7662.5

The first low sub-channel is given by substituting n = 1

$$f_1 = f_0 - 248.5 + 28n = 7662.5 - 248.5 + 28 = 7442 MHz$$

The corresponding first high sub-channel is given by

$$f'_n = f_0 - 3.5 + 28n = 7662.5 - 3.5 + 28 = 7687$$

First high sub-band – first low sub-band = TR spacing

7687 – 7442 = 245MHz

Using similar procedure the calculated channel plan is given in the table below

Appendix F: Microwave channel arrangement

Low	High
7442	7687
7470	7715
7498	7743
7526	7771
7554	7799
7582	7827
7610	7855
7638	7883

ITU-R mentions that this channel plan overlaps with the 8GHz block 7725 – 8500MHz in F.386

Printed in Great Britain
by Amazon